From Kabul To Peshawar

KHALIL RAHMANI

authorHOUSE

AuthorHouse™
1663 Liberty Drive
Bloomington, IN 47403
www.authorhouse.com
Phone: 833-262-8899

© 2020 Khalil Rahmani. All rights reserved.

No part of this book may be reproduced, stored in a retrieval system, or transmitted by any means without the written permission of the author.

Published by AuthorHouse 11/06/2020

ISBN: 978-1-6655-0227-6 (sc)
ISBN: 978-1-6655-0228-3 (hc)
ISBN: 978-1-6655-0226-9 (e)

Print information available on the last page.

Any people depicted in stock imagery provided by Getty Images are models, and such images are being used for illustrative purposes only.
Certain stock imagery © Getty Images.

This book is printed on acid-free paper.

Because of the dynamic nature of the Internet, any web addresses or links contained in this book may have changed since publication and may no longer be valid. The views expressed in this work are solely those of the author and do not necessarily reflect the views of the publisher, and the publisher hereby disclaims any responsibility for them.

Special thanks

My dear wife Roya Siddiqyan Rahmani
Kawa Shafaq Ahang
Adam Siddiq
Abdullah Ali-Ahmadi
Zalmay Rahmani
Bari Rahmani
Mahbob Wali
Kabir Rahimi
Murtaza Pardais
Fawad Akram
Sabaa Akram
Rebecca Sadeed
Elhama Sediqi
Mahshoq Rahim
Anna Zakharova

Leaders of communist parties

Leaders of Islamic parties

PREFACE

From Kabul to Peshawar is not just the story of one individual's journey, but the theme of what the entire nation of Afghanistan has been subjected to by a double-proxy war which robbed the country of both its material assets and spiritual gifts. This nation, which had so much promise for the future, had it's evolution halted when it was divided between two extreme political viewpoints. The consequences of this division took the people down to an unknown path that has caused millions of people tremendous despair over the past four decades. A nation that was run by a few ignorant, power-hungry, ill-minded sellouts who hand it a shovel to dig its own grave and it did.

The story of From Kabul to Peshawar is one person's reality. It is a firsthand experience as a result of being ripped away from his homeland and thrown into a reality in which survival was his constant battle. His journey takes him through the path of disaster between Afghanistan and Pakistan during a time of war. During this journey he comes to know the ugly truth about: this war which is a political proxy between the East and the West, between the U.S.S.R and the U.S.A. These two superpowers could care less about the millions of lives lost and an entire nation of people they left crippled and wounded. They could care less for the millions of orphans and widows they created. It's on this path that Khalil realizes the three lies of the world: communism, western humanism, and religious compassion, all of which have been wrapped in the velvet of familiar propaganda. This story is an example of how war destroys standards, deprives humans of morality and values, and creates violent tensions that corrupt and wound both the present and future generations to come.

In From Kabul to Peshawar, we also see the fate of a generation that has been marginalized from the context of society. This generation is the generation of storytellers who, from the time they were born and opened their eyes, were placed in the forefronts of the Soviet Union, the United States of America, and their allies. The structures created by this society become disrupted by the monster of war that devoured the material and moral assets of society. In such a situation, the author's generation has four options:

Option 1: Take the gun and follow radical Islam in their US-backed war against the communist regime and kill his compatriots for America's victory over the Soviet Union.

Option 2: Take the gun and follow the puppet government of the Soviet Union and kill his compatriots for the USSR victory over America.

Option 3: Leave the homeland, embrace and embark on a journey to an uncertain future.

Option 4: Be forced to stay in the land of fire and blood without any guarantees of survival or an uplifting future.

All four options make it impossible for this generation to present themselves in the context of its history, society, or to stand up for the aspirations of its own generation. In this way a generation is lost between Kabul and Peshawar, and its absence causes a cultural void and the interruption of the transmission of shared values from one generation to another. That is why this generation, our generation, has never had a pioneering role in future developments until today, because it was so disrupted that it could no longer be considered «one generation".

The author of From Kabul to Peshawar, with all his misfortunes and hardships that came his way from Afghanistan to Pakistan, is one of the few fortunate people to finally achieve and embrace his ultimate goal while thousands of other people lost their lives along the way or survived witnessing atrocities that took place along the way. From Kabul to Peshawar is written in a very intimate and unassuming language. The narrator didn't make the story a myth. Instead, he expressed his vision of the truth the U.S. and U.S.S.R. made for him. The historical question that remains is: which one of the superpowers is responsible for all of these barbaric and unimaginable crimes? The U.S., the Soviet Union, or both?

Kawa Shafaq Ahang
From Kabul to Peshawar 1980

Kabul during the civil war between Islamic parties

CHAPTER ONE

Afghanistan was once a tranquil, peaceful heaven for its citizens and more than three million tourists who visited the country each year. People were very open, friendly and neighbourly to one another, treating each other as family. It was common to walk into your neighbors' homes without any invitation. Life was about sharing, caring, loving, and enjoying everything together. Trust and cooperation were the baseline of how society functioned. It was the best time and place to be a young man or a woman. Homes were filled with playing and enjoying music. Friends got together regularly to play sports in the parks. Parties and festivities were happening weekly. It was safe to go anywhere at anytime. There was nothing to worry about or feel even the slightest bit of fear.

I remember a time when I was six years old and went to the beautiful city of Bamiyan where the famous statues of Buddha were carved into the face of a mountain. Zia, a family relative, had done a favor for a gentleman who wanted to pay it back by granting our family the use of a bus for a trip to Bamiyan[1]. Zia consulted all the family members about this trip and together it was decided that a few families would get together and make use of this opportunity for a holiday. In the bus I was excited beyond measure. Looking out the bus windows, I was mesmerized by the beautiful, picturesque, lush green towns and villages.

We arrived mid-day at a campsite located on the top of a hill. It was surrounded with breathtaking views of Bamiyan valley and the majestic statues of Buddha. The campsite had Mongolian-style yurts for each family to occupy. Between the campsite and the statues of Buddha there was a vast stretch of agricultural farming land with beautiful willow, plantain, sycamore, almond, walnut, and cherry trees across. In the morning, the sun rose and showered the statues with its rays, giving the whole place a magical and feel. A combination of natural beauties and man-made creations made the experience of being there pure miraculous.

The campsite was occupied with a mix of Afghans and international tourists. In the midst of its tranquility, there was a joyous atmosphere. We visited Bamiyan Hotel, which was solely occupied by European tourists. As for the statues, the main two are fifty meters and thirty-five meters tall. There

---

[1] Every spring we went to Mazar Sharif where the famous " Mila-gul-Surkh" was held, or "Mazar's Tulip Festival". Most winters we went to Jalalabad.

Buddha Bamiyan

Foreign tourists

are two smaller statues as well. Our elders explained to us that the two large statues are the parents and the smaller ones are their children. The mountain, which was peppered with caves, was said to be a place where monks took refuge within. These monks were the guardians of treasures stored there and those treasures were looted throughout history by cruel foreign invaders.

As we walked to the foot of the Buddha statues, we were awe-stricken by their size and overtaken by the desire to explore them further. We had a guide. He offered to take us to the top of the largest Buddha statue, where we could see the structure from within. One kilometer in, there was a path to the top of the mountain which we began to climb. As we reached the top, we came across a locked wooden gate. The guide unlocked the gate and ushered us through. A tunnel lied on the other side. Passing through the tunnel, the sun light shined through the manmade boreholes which were strategically carved to allow natural light inside.

Eventually we reached the head of Buddha. It was a circular flat space for about fifteen people to stand and see the views of the valley below. The thrill of standing on top of Buddha's statue,s head and seeing the picturesque and heavenly beauty of Bamiyan valley filled my heart with joy for all of eternity in that moment. The mountains, hills, bazaars, buildings, houses, trees, and lush green farms were captured by my eyes in form of an unforgettable memory. From elders to children, we were all mesmerized with. Nobody wanted to leave. The guide had to eventually nudge us to leave. We had a wonderful day in Bamiyan city.

The next day we set off for Band Amir[2]. The road conditions were bad unpaved roads with no signs and strong winds. As the bus was meandering through valleys, singing, dancing, and laughter filled the atmosphere inside. Suddenly, the car came to an abrupt halt. The driver told us the bus was stuck in mud. Everyone pushed the bus as the driver gassed the pedal. Unfortunately, the bus kept sinking deeper in the mud. We were helpless.

---

[2] Band-e Amir National Park a national park located in the Bamyan Province of central Afghanistan. It is a series of six deep blue lakes separated by natural dams made of travertine, a mineral deposit. The lakes are situated in the Hindu Kush mountains at approximately 3000 m of elevation, west of the famous Buddhas of Bamiyan. They were created by the carbon dioxide rich water oozing out of the faults and fractures to deposit calcium carbonate precipitate in the form of travertine walls that today store the water of these lakes. Band-e Amir is one of the few rare natural lakes in the world which are created by travertine systems. The site of Band-e Amir has been described as Afghanistan's Grand Canyon, and draws thousands of tourists a year. The river is part of the system of the Balkh River)

Foreign tourists

We decided it was best to wait inside. Our only hope was that the authorities whom we told about our journey would come searching for us after they realized we did not turn up at our destination. Fortunately, that was the case. At 7 o'clock in the evening, an army truck came and pulled the bus out of the mud. We were all very hungry and exhausted, so they took us to the nearest café. The café owner was very kind and generous. He served us with an abundance of food. At this great feast, I had my first experience of eating the famously known dish called "chaaynakey", an Afghan soup dish made in teapots. Night had already fallen, so we made our way swiftly back to the Bamiyan hotel after our meal.

A crisp morning breeze awaited us. The authorities advised us to take a different route to Band Amir. As we neared our destination, my eyes were wide in amazement, staring at the unbelievably beautiful natural lakes and their deep blue colors surrounded by impeccably gorgeous mountainous terrain. As the bus came to a halt, I couldn't contain my excitement. I ran out the bus to have a better look at this majestic beauty. We spent the whole day by the lake until the sun began to set. Boarding the bus, we all took a few glances back to the deep blue waters before driving towards Kabul.

As we were on our way to Kabul, late in the night in middle of nowhere, we came across a young European woman who was hitchhiking through the country by herself. Our bus stopped to give her a ride. She was a very jolly character, great in spirit. We gave her some food and those of us who understood English spoke with her. My mind was racing with curiosity. How did she end up in the middle of nowhere in Afghanistan by herself? What I didn't know was that large numbers of Europeans who called themselves young hippies hitchhiked from Europe to Afghanistan and beyond. The level of safety in Afghanistan was incredible. A complete foreigner who knew nothing about the culture or language could travel safely anywhere in the country, even in the most remote corners. Just like this young European girl, there were countless women roaming around the beautiful country enjoying what each province had to offer and having a marvelous time. Every house, every street, and every city of the country was filled with joy, love, laughter, happiness, and safety. Every family was more than pleased to offer whatever they could to others and especially to foreigners.

Foreign tourists

Foreign tourists

CHAPTER TWO

Our house was built in a way that the beige-colored, two-story building was directly parallel and facing the street, and the backyard that was mostly covered by concrete was situated behind it. The backyard consisted of a small lawn area that was watered almost every late afternoon by an electric water pump that was connected to our water well. My sister Soraya, who at one time studied in Russia, brought the electric water pump with her from there. None of our neighbors in the area had such a pump. Instead, they would generally use a bucket that was tied to a rope to get water from the well, which was not easy and certainly not a desirable method. Thus, our neighbors would use our water pump, specially since after a couple of buckets, the water would be cold and taste so crisp as if it came directly from the mountains and valleys of Paghman or Salang.(The rivers of both cities have cold and clean water)

My father had tastefully planted various kinds of flowers around the lawn which had fully bloomed and were adorning our little backyard. One corner of the lawn had a peach tree which did not bear any fruits. In another section of the backyard, five grapevines had been planted. They were covering a pergola that was beautifully put together by a very skilled carpenter of the neighborhood. The grapevines were attached as high as the top of the wall and grapes were hanging from them like shining chandeliers, inviting everyone to eat from them. Intertwined with the pergola, every visitor commented about how the grapevines beautified our small backyard. Although the grapes our vines bore were small and known as 'keshmeshi', in terms of appearance, their tenderness, sweetness were unparalleled to any other grape I've eaten. At times, some grapes that were covered by leaves, only to be discovered late in the fall season. Those ones were the most delicious surprises to find. Our neighbors also enjoyed our grapes.

The walls close to the grapevines were covered by paintings of my eldest brother, Zalmay Rahmani, who was a well-known artist at the National Theater. The paintings mainly depicted love scenes would combine with the lovely colors of the leaves during the fall season, creating an incredibly beautiful scenery. In fact, my brother's artistry could be seen on most of the walls throughout our house, invoking amazement and admiration in spectators.

In the other corner of the yard, there was a separate building with two

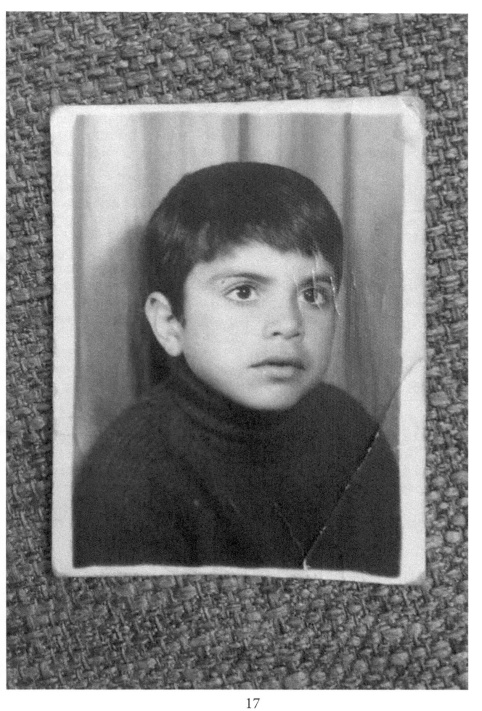

rooms, toilet, and a kitchen. The tall walls that separated the dwelling from the road created privacy for us. The spirit of freedom was joyously flying throughout our house and my father supported this freedom. In religious society the head of the house determines the type of life and the ruling culture of the home. My father ensured freedom of expression was a culture we all enjoyed in our home. Although my grandfather and mother were very strongly affiliated with religion, they also gave us lessons about the customs of ethics, respect, affection, and love of others. Joy and laughter were the key hallmarks of our family.

We are a family of five brothers and two sisters, my elder brothers Zalmay, Farid, two sisters Soraya, Rohafza, two brothers Bari and Zabi, with me being the youngest child of my family. Growing up as the youngest family member, I was at the center of attention and kindness from everyone. I was a jovial, lively, and a witty character. My character was infectious to people of all ages. Thus made me popular to everyone. Most of my friends were girls and women, as well elders of the neighborhood who'd often invite me into their homes and prepare delicious foods and drink for me. In return I would make them laugh. I also had adversaries and enemies with whom some conversations often led to complaints and in some cases even to one-on-one fighting. Shortly after, we would make up and be friends again. Overall, I was a rather social child which in my opinion is the key to friendship and popularity.

Guests were allowed to our home with open heart. Just as the boys were allowed to live freely, my sisters also had freedom, and the issue of gender was not at all questionable. Thus oldest sister Soraya used this opportunity to get a scholarship and study in the Soviet Union. Unlike thousands of other Afghan families, as teenagers in Afghanistan, we were spoiled. Being told by our parents and others that we couldn't do any work like, never less having the capability of going to the grocery store to get food.

Although our cousins always gathered at our house, there were many friends of sisters and brothers as well all times in our house. Our home was filled with music, singing, joking, and laughter . In those times, having a phone was a symbol of being rich. Coming from a modern family we were the only ones among relatives to have a landline in our home. The price of having a phone was one that not many families could afford. This new innovation marked an age of curiosity in Afghanistan, bringing family

Me with Middle school friends

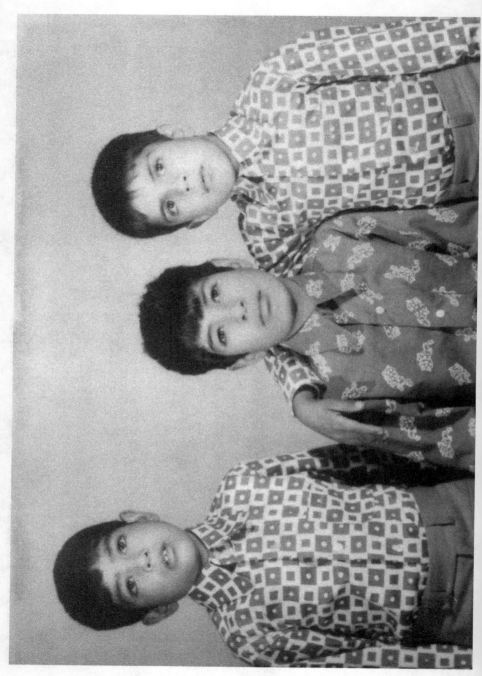

Me with my brothers Zabi and Bari

and friends from far to see and hear what the phone sounded like. During gatherings when the phone rang, friends and family members would run to be the first to pick it up and experience it first hand at our home.

I was in love with Kabul,s four seasons. In spring, the flowers bloomed, the air was filled with the smell of fruit trees and the rain. It was also the time for schools to start and the street were filled with joyful students. Spring was the time for festivities such as the festival of Roses and Tulips.

Summer was the break time for students. It was the season to enjoy delicious fruits. For several weeks people would celebrate Afghanistan's Independence Day and the entire city would be bright with lights. Kids and the youth would enjoy their time in one of the many swimming pools around the city such as Makroyan Pool, which was for women only every Tuesday of the week, Pul-e-Charkhi and Intercontinental were coed pools that allowed men, women, boys and girls to swim together. Other swimming pools included Silo, Bagh-e-Babor, and Club-e-Askari (Army club pool). Summer time was also the season when my older brothers Zalmay and Farid would go partying in one of the many night clubs of the city with their girlfriends. Some of the night clubs of that time were Golden Lotus, Khoum-e-Zargar, Golden Cave, Club 25, Club 99, Chinchila, Golden Lily, and Maxim. The young generation would party, drink, dance and enjoy themselves.

Fall was the season of colors. Fall season was known for the sounds of colorful leaves falling from trees and the season for delicious pomegranates.
Students were enjoying the last days of school and looking forward to their winter break.

Winter was filled with beautiful white powdery snow. Sometime, it would snow for weeks. Winter was the time to stay indoors. It was time to enjoy all the delicious dried fruits while reading suspenseful books or listening to stories on the radio. It was time for flying kites or going to winter school. It was time to spend the nights listening to wonderful stories of our grandparents.

During the summer when schools were off, my father would rent a house in a city called Paghman. The area was rich with green trees full of fruits, lakes, and beautiful weather. In addition, it had a breathtaking view from every angle of the mountains and city. During these times, we were mostly

with family that would travel with us. In winter my father would rent another house in a city toward East Kabul near Pakistan called Jalalabad. a very famous city known for its majestic beauty and weather.

It is said that in our house, music was performed intermittently, by famous singers such as Sarban as well as up and coming new artists.

My elder brother, Zalmay Rahmani, who was heavily influenced by Mr. Sarban's fame desired to become a singer in order to become famous and popular, but since he was not talented in singing and composing, he tried in the field of art.

After seeing a few performances with friends, he was encouraged to further his talents in acting and enrolled in Onara Zeeba (Fine Arts) School which he graduated from among many other notable singers, actors, and artists. When Faiz Mohammad Khairzadeh returns home after graduating from the US Department of Cinema and Theater, he used his financial resources, and influence, to create an institution called Fine Arts known as Onara Zeeba . Of course, the institute was by no means was within the framework of Kabul University and higher education. My brother, with his small talent in acting, would later be accepted in this institute as an actor in theater and cinema. The institute had two departments music and art. The artists who were trained in the institutes later became famous. Some known artists include Zahir Howaida, Rahim Jahani, Aziz Ashna, Fouad violin, Chitram, Mazideh Sooro, Khan Agha Sooro, Yahya Shoorangiz, Akbar Ramesh, Zahir Qaumi and a few others. In the allegory section (Art), Huma Ebrahimi, Meymoneh Ghazal, Mahbooba Jabbari, Zuleykha Negah, Zarghoneh Hakimi, Moghaddasa Hidden, Eghlima Hidden, Gol Maki Fataneh, Ehsan Atil, Karim Javid, Majid Ghiasi, Sattar Jafaei, Faghir Nabari Khoshim, Nasim, Farooq Ramesh, Zalmay Rahmani and others.

The institute, owned by the Ministry of Corrections and Culture, later continued its artistic activities by changing the name of the directorate to Saghaft and Onara that was awarded to graduates of the Bachelor's Degree Art Courses. When the Institute of Fine Arts released its first artwork, an art film called Manand e Oqab (Like an Eagle), my brother played the lead role. The movie became a "blockbuster hit" and played at the Park Cinema.

From the point of view of capital and assets, we lived a modest life governed by modern thinking in the modern world. For instance, when it

Me, mom, grandma, Zalmay and new born Nephew Massi

came to celebrating the anniversary of my grand sister Soraya's birthday, the whole house was lightened.

The young ladies and gentlemen wore western makeup, hairstyles, and outfits. They danced throughout the hours of the night into the dawn to modern music. Sometimes we partied to live music. Twice, Naim Popal, an innovative artist in Afghan pop, came and performed with his band at our home. The neighbors watched and enjoyed from roofs and walls. We had a very friendly relationship with all the neighbors without exception, and there was mutual respect between us.

The room of one of the brothers was so well decorated that it was surprising and interesting for everyone. Each millimeter of the room was perfectly designed, giving it a reputation in the family and friends circle.

Upon entering Farid's room, which was the largest of all the rooms in our house, you saw that he divided it into several sections. I have to describe Farid's room in the following order: the left-hand corner, the right-hand corner, and the center of the room. The right side of the whole wall was adorned with large posters of Googoosh (Iranian singer and actress), each one more beautiful than the last. This meant that most of the more than fifteen photos of Googoosh were especially impressive in a unique way.

Every corner of his room, all the way to the ceiling, had extensive photographs of famous American films. One group of posters showed three photos of three different shooting modes of Western films, with scenes of duels or people shooting at each other. The bullets, and more recently the bloodshed, were the result of the film. He had arranged these photos, buying them either from the shop of Khairzadeh magazine, located in front of Cinema Park, or with his good public relations; he took it from Suliman, the director of Cinema Park.

Next to the right-hand side of the room was a large glass door covered with a stunning curtain. If I'm not mistaken, the red velvet was covered. The window faced the neighborhood street, and sometimes from the balcony, you could see the incidents that occurred in the neighborhood. Next to the window, there was an alcohol beverage bar in the corner of the room that was his own handmade. Around it, he covered it beautifully with burlap. Behind it, which consisted of two floors, you could drink whatever you wanted.

Me, Zabi, Rohafza, mom and dad, Zalmays wife, Zalmay and Bar

My sister Soraya birthday party

In front of the wall of the right-hand corner hung large and colorful photos of Charles Brownson, Alain Delon, Bruce Lee, Steve McQueen, Anthony Quinn, and others. Above these photos, he had arranged his collection of knives and various daggers with beautiful decorations on the wall on both sides of these daggers. Two old antique whips: one large, wrapped in the shape of a snake, and the other smaller.

With a little degree of distance, he showcased a collection of beautiful lighters, the most exceptional one playing music when it opened. And in the middle of which an old rifle from the time of the Afghan-British war hung with all its handmade delicacy. On the ceiling in the middle of the room, various types of rosaries hung, the most beautiful of which was the stone rosary of Shah Maghsoud Foroozeh. Next to this wall was a giant sofa that he handmade. His design was more of a style of furniture that accommodated friend's gatherings. In the corner of the wall, there was a vast curtain which he made himself out of bamboo of different colors, separating the bedroom from the living room. Underneath it, there was an Afghan double-breasted mattress with old-fashioned pillows. But what should I say about the bedroom? When I recall it, all my youth reappears in front of my eyes. His room was stunning and imaginative for us youths. First of all, the bedroom, or this part of his room, consisted of a long bed frame with beautiful pillows and a blue velvet bedsheet. In front of his bed was a large mirror that opened on both sides, beneath which were various perfumes and colognes. All the walls displayed pictures of beautiful nude models and cultural adventures in different sexy poses—the more interesting facet, which was his invention for this room. There was a massive switchboard behind his bed that controlled all the lighting. The switchboard had more than fifteen buttons that controlled the lights, which were in every corner of the room and featured different colors. Let's not forget that there were hundreds of photographs of Hollywood legends and celebrities hanging from the roof of his room, which he designed, either separated or attached. In memory of my brother. and admiration in spectators.

My sister Soraya birthday party

Rohafza and Massi in our backyard snowy days

Soraya And Farid

«کجدی قروت» دیدن کرد و کار هنرمندان آنرا ستود

Zalmay with his theater crew

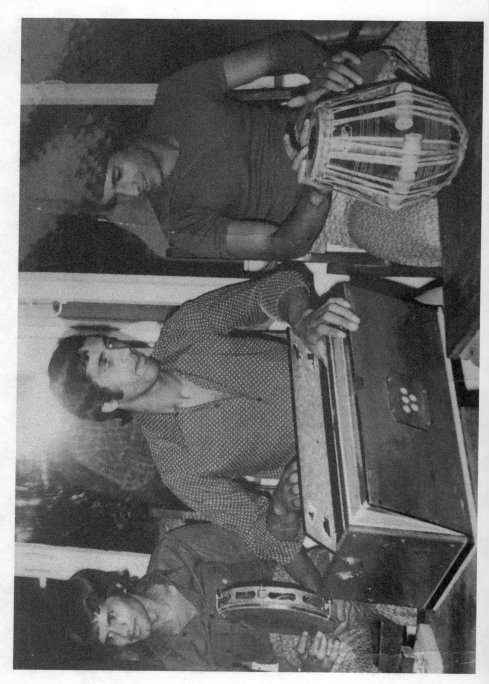

My cousins playing music

Farid in his room

Farid and Zalmay in Farid's room

Bari in Farid's room

Our father in Farid's room

Soraya Departure date to Russia. Me some other Family member, cousins and Naim Popal

CHAPTER THREE

In 1978, after the communist regime came to power, particularly after the Soviet Invasion of Afghanistan on Christmas Eve of 1979, Afghanistan became an increasingly dangerous place. Intellectuals, influencers, professionals, clergy, academics and civil servants whom did not belong to the two communist parties of 'Khalqi' (Masses) and Parchami (Flag) were imprisoned, tortured, killed or forced out of the country. It was not even safe for very young Afghans. After all, the Khalqis and Parchamis destroyed the peace. For the first time in my life, the colorful atmosphere of the motherland turned gray.

After the Soviet invasion in our homeland, my sister Soraya, who was studying in Russia, fled after her studies for Italy and eventually found refuge in Germany as a political asylum. One of my brothers Bari was called up to the army without his will. After a while, he deserted the army, left home, and fled to Iran. Another brother Zabi was imprisoned for raising the sound of liberty during the school demonstration against the Soviet occupation. A beautiful and brave girl, also named Soraya, was my classmate throughout primary school. She brought the information of brother's imprisonment to the family. After a few months, he was freed, but he was still harassed by members of Khaad (Afghan Government's Secret Service), so he also fled to Iran. My elder brother Zalmay was under the supervision surveillance of Khaad. At the time, he and his newlywed wife and new born baby Massi were hiding in with relatives and friends' homes. Unfortunately, they were eventually found and captured after a year. The communist government sentenced him to 18 years in prison. I was once arrested for propaganda against the government for a day in the party district. Since then, I had no feelings of relaxation or interest in school

Father worked in Kandahar for a living. There were no more gatherings at home. The songs of The Beatles, Elvis Presley, Googosh (Persian female singer) and Ahmad Zahir (Afghan pop singer) were not heard from our home any more either. Laughs were replaced with cries, and instead of going to parks, café's, parties, and the cinema, we were visiting the Pull-e-Charkhi Prison (the biggest prison in Kabul) to seek the rare chance of seeing my brother. As my nephew Massi became three-years-old, he asked for his father. We'd say to him that his father was on a business trip, and so the child was constantly expecting him. Our community became smaller and smaller except for very close relatives and friends. Others were afraid of being close to us.

So, we made new friends. Of these, there were two sisters and mother of Hazrat Sibghatullah Mojaddedi (Sibghatullah Mojaddedi, who served as the President after the fall of Mohammad Najibullah's government in April 1992. He was also the founder of the Afghan National Liberation Front).

Exponentially, the situation in our country became worse. I have to leave Afghanistan too. The sisters of Sibghatullah Mojaddedi who were like family members always gave me hope by promising to send me safely to Pakistan.

The government noticed that young people were escaping from the homeland or joining the Mujahideen, posing a shortage of soldiers for the army. They had organized search parties and ordered that any young person who had no school ID-card would be forced to join the army, regardless of their age. The surveillance troops, under the order of the government, attacked the neighborhoods, districts, and cities. They captured and dragged children from as young as eleven-years-old to adults as old as seventy-years-old to the army barracks. They put them in military vehicles and sent them straight to battlefield, majority of whom never came back with families expecting and waiting with hope for their fathers, wives for their husbands, sisters for their brothers, parents for their sons.

When I was sixteen-years-old, my mother decided to send me to Kandahar so that my father could arrange to send me to Iran to join my brothers. We informed those from our family and friends' circle whom we trusted of our intention to leave Kabul and shared with them to be on the lookout for a trustworthy smuggler. Before the regime change, there was a very high level of friendship, trust, sympathy, and solidarity among neighbors. Knowing and visiting among neighbors even as far as ten streets away was very common. With the advent of the new regime of the People's Democratic Party, which consisted of the political parties of Parchami (the Party of the Flag) and Khalqi (People's Party), the door for the art of music was opened for all those who were interested in singing and playing musical instruments. Amongst our circle, that included Wais, Khwaja Hamid, Hamed Parwais, and the rest of our friends who spent once or twice a week playing music and partying all night. On some nights, the late Sarbaan's son (famous Afghan singer) Abdul would join us too.

Two brothers named Zamaray and Padshah lived one street down from our house. Zamaray was about two years older than me and Padshah was probably a year younger than me. My friendship with the two brothers was more of a street acquaintance which often meant we would meet and engage in small talk and that was it. However, Zamaray was good friends with Hamed, Khwaja Hamid, and Wais. Zamaray was a big fan of Nashenas' (Afghan singer) voice and would sing like him. On some nights, Wais and Khwaja Hamid would encourage him to sing. Since they were good friends, they found out that Zamaray's sister's husband, Ismail, was traveling between Afghanistan and Iran. He said Ismail would help me travel to Iran. Since that was Ismail's occupation and since he turned out to be trustworthy, Khwaja Hamid Ahrary also decided to join me on my escape to Iran as well.

Thus, my intention to flee Kabul got serious. One night, I found my dear aunt (my father's younger sister) asleep at our house. The next morning, she accompanied me and Khwaja Hamid Ahrary to the Kandahar and Herat bus stations where we met Ismail, the smuggler. He had put a group together where I met my two friends, Wahid, famous for Laff (being mendacious) and Hayat, the son of Janat Khan Gharwal (The head of a commercial bank called Pashtani Commercial Bank, and he built a mosque in our area the name of the mosque was Gharwal Mosque. He was a supporter. Of King Zahir Shah.) by seeing them I became courageous and confident. Hayat was my classmate in seventh and eighth grade. Our friendship was just like the bond of two brothers.

We finally arrived in Kandahar and settled in a hotel. Kandahar was a vibrant city. Restaurants and cafes were open and the streets were crowded. The city was established by Ahmad Shah Durani in the early 18th century under the name Ashraf-Al-Balad and it was declared the Capital of the State of Afghanistan. Due to the existent Eid-Gah (place of the two Eid prayers) and the cloak of the Prophet of Islam, the city was considered to be amongst the holiest cities of Afghanistan. It is a common belief that because they are residents of such a joyful and happy place with the city brimming with lively restaurants, cafes, and the sounds of local folk music and songs from Bollywood and Pakistani movies that are played back through record or cassette players, or other special music systems, can be heard loudly from afar. In the same way, the city's only movie theater that mostly showed Bollywood movies and often movies with many fighting scenes, was full of spectators and movie fans. At times when all of the tickets would be sold out and no more seats available, some of them would go as far as to bribe the theater authorities to let them watch an entire three-hour-long Bollywood movie while standing. Eating, drinking, smoking of cigarettes, or even hashish was a common occurrence inside the movie theater in Kandahar. Kandahari fruits and ferni (milk-based dessert) that one could buy street-side are very delicious.

My father came to me every day after he finished work. I told him that the smuggler knew of only one route to Iran. It was just our bad luck that it was blocked because of the war and nobody knew when it would reopen. After two weeks of waiting, we had to come back to Kabul. Ismail had already been paid for my journey, therefore, he would either send me to Iran or return the money. Well, it was decided that we should wait until spring.

With the passage of each day, the situation in the country continued to grow worse. One day, the smuggler came to take me to Iran. Khwaja Hamid and Whaid Laff both rejected the trip and no longer wanted to leave the homeland. This time, Hayat, my best friend, and I were on this trip. The smuggler said he had to send us one day earlier than planned to Kandahar, stating it was for security reasons. Hayat slept at our house that night. The next morning, we left to the bus station for Kandahar. As we arrived in Kandahar, we noticed that it was no longer the Kandahar of a few months ago. The walls were full of bullet holes. The roads were damaged and empty. Most of the shops were closed and abandoned. There were only soldiers, military jeeps, and tanks to be seen in every corner of the streets. We stayed in the Kabuli hotel. The bus driver dropped us far from the hotel area. As we crossed the road, a soldier saw us and beckoned us to him. He asked who we were and where are we going, we anticipated this and our answer was ready: "My father works here and we have not heard from him for a long time, so we came to here to ask for his well-being." The soldier sked for our National IDs. On my NID, there was a picture was from my childhood which made him feel suspicious towards me. He said: "You are using your younger brothers ID and you are an Army deserter!" Hearing this, our lips became dry and no blood seemed to be running in our veins, I didn't know what to say to him now. As luck had it, the commander called the soldier for an urgent matter on the other side of the square. The soldier gave our NIDs to another soldier and rushed towards his commander. The second soldier asked us with a very friendly smile if we were boys from Kabul. We replied, «Yes.» He said, "It's not safe to be here for any destination. You two need to get out of Kandahar as soon as possible. Take your NIDs and run away, now." We left for the same hotel.

After sunset and all the way through the night until sunrise, the fires started and the bullets lit the town's night. We did not sleep at all. I told Hayat if the smuggler could not come with a plan or lie to us again, we cannot stay in Kandahar as it is and that we must return back to Kabul. My premonition proved right and we ended back in Kabul.

Some weeks later, Hayat left for Pakistan. I waited until summer to attempt at going back to Kandahar. Summer came and again I was on route to Kandahar. We were not far from Ghazni. Our bus was stopped by a group of Mujahideen. This was the first time I saw Mujahideen. Among all the passengers, they singled me out and took me to their commander across the road. On the side of the road, there was a stream full of water, and on the other side of the stream there was the Mujahideen commander. He started interrogating me in Pashto language, "Who are you? Are you a communist or a spy? Where are you going?"

My answer was the same: «My dad works as a Chief Executive in Kandahar and we have not heard from him for a long time.» He asked me if I am a Muslim. I replied, "Yes." Then he said, "You should recite the morning prayer." I read the only bit of Quran I knew, which is al-Humadu and three of Qull for him. He asked again to pray morning prayers for him. I repeated him an Al-Hamdu and three Qull, (the opening and the last three Surahs of the Quran. Because they are short but powerful, they are generally the first verses of the Quran that Muslims memorize to meet the minimum requirement for the five daily prayers). I memorized the same thing in Arabic only and I still remember them. the commander slapped me and I almost fell into the water. He said that the communist infidels do not know the prayers. An elderly white-bearded man approached us and said to the commander, «This child doesn't know the meaning of communist. He could have forgotten the prayers out of fear. His neighbor sitting in the bus and he is saying that the boy is actually going to his father, that the family not heard from his father a long time." The commander accepted the old man's words and abandoned me, but said with anger, «Make sure to remember the prayer!»

We sat in bus again and left towards Kandahar. A few kilometers away from Kandahar, government soldiers stopped the bus and asked some passengers and myself to get out of the bus. This time, Ismail, the smuggler, was brave and came down with me.

The soldiers led us into a squadron in the commander's room. Again, the same questions started: who are you and where are you going? My answers were also the same as always: My father is in Kandahar and because we have not heard from him, my whole family is worried and I came down to see him. He looked my NID up and down. He asked one or two more questions from the smugglers then we were allowed to leave with dignity. Back we went onto the main road.

Again, the soldiers stopped our bus at the suburbs of Kandahar, they told us this is the last stop and we all must alight the bus. We had to walk to a nearby district/village, and we checked in a hotel far from Kandahar city. That same evening, my father came to see me. This time the smuggler decided to enter Iran via Herat. The next morning, we took the bus and drove to Herat, we were just a few 5- kilometers away from Kandahar that the soldiers stopped our bus and once again me or two others were asked to come out of the bus. They didn't allow the Ismail smuggler to come down from bus neither did he himself wants to accompany me. He knew that I won't be allowed back, quickly he mumbled the name of a hotel in Herat. Just told me which hotel I could visit him in Herat.

The soldiers took me into the military quarters. In the center, there was a building with two rooms. They brought me inside that building. When I got in, I saw that both rooms were full of men and young boys. I noticed a few familiar faces from the Kabul. They were all like me, wishing to flee to somewhere safe from the communist atrocities, and unfortunately, they were arrested and waiting for their destiny to be decided by these soldiers. I spoke to one of the familiar faces to find out what the situation was here. He told me that the soldiers captured everyone from various vehicles and now it's suggested that they are going to be sending us back to Kabul.

While talking to the boy, a soldier from Hazara tribe entered the room. This soldier was small in height and his uniform was two sizes larger than his actual size. He looked around the room and directly pointed to me. He told me that he knows me from Taimani (my hometown district of Kabul) because he was from Taimani himself. I asked why he joined the army in Kandahar. He answered that he was going to Iran and was captured here then forcibly recruited in the army. Then he paused and spoke to me silently about how he is waiting for the right time to escape and go to Iran.

I begged him to keep my father informed and with him being a kind and generous person, he agreed to send my message to Father.

It was nearly dinner time that my father came with permission from a high-ranking government official and took me with him. I insisted on him leaving me here because the government itself will send us all back to Kabul. When we left the military camp, my dad told me that it was a lie, that they are sending those kids to the front line of the war. My dad advised me against travelling to Herat because the road was full of dangers, finding the smuggler would be very difficult, and it is not known whether he is in Herat or not. He suggested I should fly back to Kabul. I willingly agreed with my father because I was scared of going to Kabul by bus.

I had to wait five days to fly. My father was hired by the Ministry of Public Works in Kandahar in a hotel project at the accounting department. The project had a Baba (old man) gatekeeper and he had nothing else to offer except telling endless stories, one after other. Baba had visited Kabul a couple of times and was telling stories of cleanness and the beautiful people of Kabul. I used to listen in depth. The way Baba talked about Kabul made it sound like a foreign city to me, as if I had never been there. Baba was particularly fascinated by the district of 'Makroyan and the ladies who came to the balconies. (Makroyan is located in the northeastern part of the city). Makroyan was one of the first multilevel occupancy buildings built in 1960s with the help of USSR during the King Zahir Khan era. However, the distribution of the Makroyan residences happened in the Dawood Khan era.

Those who studied in USSR or had a Russian spouse were among the first to be offered these residences. Government workers were next in line to have access to these residences. Makroyan residences had balconies and were very modern for the time and included well equipped bathrooms with showers having access to hot and cold water. Initially, the people of Kabul disliked Makroyan and were not too enthusiastic about apartment living. It was one of those days, after hearing all the beautiful stories of Baba, a young local man sighed and said, "Regardless of whatever happened, I will visit Kabul this year." Five days later, I flew to Kabul, which was the first time in my life flying on an airplane.

CHAPTER FOUR

The smuggler never returned the four thousand Afghani (Afghan currency) and asked repeatedly to go to Iran with him, but I no longer had the interest nor dared to go to Iran with him. One of my uncles' (my mother older brother) wife's family members knew the way to Iran and we decided to go with him along with my uncle's family. Everything was set for us to go together, but one day my grandmother came to my mother and told us that my uncle's family has set off for Iran and they decided to leave me behind.

My brothers traveled from Iran to Germany and this made me feel upset and lost. They were the reason I was going to Iran. From then on everyone advised me to go to Pakistan. Sibghatullah Mojaddedi was a prominent anti-government leader and he was guiding the war from Pakistan while living like royalty over there. As it so happened, his family were very close friends with our family. His sisters and mother of promised me and my mother that they would take me with them to Pakistan as soon as possible. Again, one day I heard that the sisters of Sibghatullah Asma and Atiqa reached Pakistan without me. It was so disappointing. Everyone was distressed about me and I was a carefree young man with no real understanding of the situation in Kabul and its potential dangers, just trying to enjoy every day in the city with friends.

During this period, four new developments happened in my life: cigarette smoking, drinking liquor, the first kiss of a girlfriend, and for a few months, a tendency toward Islam...to the extent that I even wrote to my brothers in Germany, preaching them Islam through my letters. They are still making jokes of me, reminding me of my Islamic reaction. The reason for this development was the influence of my circle of friends.

The winter of 1980 was a very cold one for Afghanistan. One winter night in the month of January, I returned home from my outings, my parents were waiting for me, and they told me that there is a man that would take me to Pakistan tomorrow. That night, before the curfew that lasted from 10 p.m. to 5 a.m., I was told that we would reach 'Makroyan. They told me that the man is a family friend of my father's aunt. We always had a very good relationship with my father's aunt, and this was especially reinforced by barbarity of Khalqi and Parchami and the suffering that both our families were inflicted by them. My father's two aunts' two young sons were arrested and executed by Khalqis and Parchamis. One of them is the late Latif Mahmoudi and his cousin Dean Mohammad Mahmoudi (brother of the popular Afghan singer Sarban).

I grabbed some clothing, I do not remember what exactly, and placed it in a small bag with five-thousand Afghanis, of which I kept five-hundred Afghans in my pocket and sew the rest of the money in my underwear. The clothing for my trip consisted of two socks, plimsolls for shoes(sneakers), underwear, and an Afghan spring traditional clothes, a leather jacket which was sent by my brothers from Germany, and a cotton-padded coat.

We took a taxi with my parents and my three-year old nephew Massi to my father's aunt's house in Makroyan. A man who was about thirty years old was waiting for us at the guesthouse. My dad's aunt's daughter introduced him to us and said, «Wali is a very good friend of mine and knows the way to Pakistan and can safely take Khalil to Pakistan.» Later, she talked to Wali and he reassured my parents that he knows the way to Pakistan like the palm of his hand and he promised that he won't let any harm come to me. The man was calm and relaxed. If you did not ask, he wouldn't talk. He could be trusted. My mother was certain that Sibghtullah's sisters would help me in Pakistan. She said to Wali, "First of all, it's the will of Allah, and then it's yours. Just drop Khalil at Sibghatullah's and they will help him."

Marzia, my uncle's daughter, who was my peer and a good friend, arrived at my father's aunt's house at night. We all stayed overnight in Makroyan. The next morning after breakfast, I said farewell to my family and set off for Pakistan. Everyone, especially my nephew Massi who wanted to go with me, was crying. It felt as if it was my last farewell to my family and none of them would see me ever again. I knew I would never return again to Kabul this time and there was no longer any hope for a visit with them here again.

Near the Ministry of Finance and behind Aryana cinema, there was the bus station to Jalalabad, Torkham, and Dehsabz. The town Dehsabz was where we started the journey to Pakistan. Near Dehsabz, Wali told me, "Cousin! the fact is that I have been to Pakistan through Torkham once or twice and because of my utter respect to your aunt's family, I couldn't refuse when your aunt's daughter asked me to get you to Pakistan with me. But don't worry. I know a man who is like a brother to me. His name is Nazir and he knows all Dehsabz. He's a teacher by profession and he'll send you to Pakistan safely."

We arrived at Nazir's house. The surrounding walls were so tall that the house itself was not visible from the outside. A young boy opened the gate. Wali asked him, "Is your father home?" He shyly replied, "Yes." He called for his father, "Agha! Agha!" After a while, a man full of brute hair, tall, and a very serious face came to the gate. He smiled to Wali enthusiastically, as if he was expecting him. He hugged Wali and shook hands with me. He said to his son, "Go and tell the women to move aside." We entered the guest house. The four sides of the guest house were furnished by mattresses having one or two pillows that were designed with carpet covers over them. After a short while. tea and sweets were left behind by the door. Nazir brought the tea in and the tales began. I barely listened to their words and they didn't make any sense to me anyway. I was lost in my thoughts and imaginations the memory of my nephew crying was haunting me.

A middle-aged man entered the room. I don't know what his relation to Nazir was, but they knew each other very well. The dark of the night was spreading slowly. A lamp was hanging in the middle of the room. The whole house was silent. There were no radios, televisions, or any other kind of electric devices to be seen anywhere. Other than the words of Wali, Nazir, and that third man, there was no other sound I could hear throughout the entire home. It was as if the house just had this one room and no one else lived there. Then, I heard the dogs start barking from outside. To go to the bathroom, we would have to go through the yard of the house. Dinner was ready.

They spread a bedsheet as a tablecloth on the floor. The same little boy brought a water pot and basin for us to wash our hands. One towel was given for everyone and they began to bring the food, walking over the sheets to put the plates in front of us. After the dinner, Wali said to Nazir, «Khalil is one of the relatives of Hazrat Sibghatullah Mojaddedi and he wants to go to Pakistan. Can you help him to reach Pakistan? His family is very close to us." Nazir looked at me and said, "I'm disillusioned by all the groups and fighting." He went pensive and just looked at me and Wali while thinking and mumbling absent-mindedly, picking the threads of carpet and twisting them with his fingers. "I don't care if they are good or bad. I have lost my younger brother in their selfish war. He was only 19 years old. As you are relative of Hazrat Sahib, I will take you to the

mosque tomorrow morning. There we will find someone who may help you." We slept late that night. The lights went off but the dog's barking lasted all night. Because of the fear of darkness and dogs, I had to use the toilet and held my urine in my bladder until the morning.

The next day was a very bright and mild winter day. After breakfast, we walked towards the mosque at the center of the village. Everyone greeted Nazir along the way from his home to the mosque. We took out our shoes and got into the mosque. When we entered the mosque, a number of men stood up to honor Nazir. This was the second time that I was ever inside a mosque. The first time was in Kabul's famous Shah-do-Shamshira mosque. Inside this dark mosque, the men were sitting around with their backs to the walls. There were about 20 men. At the top, three white-bearded men were sitting. Nazir went to the three white bearded men and hugged them. We sat among these men. Nazir told them that I am a relative of the Hazrat Sahib and I that I was going to Pakistan. One of these white-bearded men said, "We don't know who is going from here to Pakistan nor when. You should go to Committee Jabha-e-Milli Nejat (the name of Hazrat Sibghatullah Mojaddedi party), because from there many Muslims are going.» He gave me the paper and pen and said, "Write: From Mullah ... (I have forgotten his name). I am one of the relatives of Hazrat Sahib and want to go to Pakistan. Give this letter to Mullah Mohammad». My handwriting was so bad and difficult to read. It was easier to read a doctor's prescription than to understand my handwriting. Nonetheless, I wrote what he dictated and then he put his stamp on it.

Someone told me something in Pashto and Nazir turned to me and said, «It's lucky that a few minutes ago, a young man left on his way to the committee. Go along with him without any worries.» I said my farewells to Nazir and Wali. I joined two boys who were my age and moved towards the Safi mountains.

As we reached the bottom of the Safi Mountain, a child was climbing the mountain. The two boys called his name repeatedly, but their shouts were not heard by the boy. After persistent and noise, finally he heard our voice and looked at us. The two kids told me, «Go ahead with him. If you do not keep up with him, just follow the footpath. That will bring you to the committee". I looked at the footsteps of people had been walking for a long time on a straight path and I started climbing the mountain. Once again, I saw the kid, but I did not see him after that and then I was lost. I was climbing like a mountain goat crawling, putting my hands and feet on stones. I was anxious to join that kid or else I would get even more lost. He was my only hope in a vast, empty mountain range.

Finally, I reached the peak of Safi mountain. I felt dizzy. My lips were dry and thirsty. I had no more strength in my legs. I climbed the entire mountain with the hope of reaching that kid, not resting for a moment. When I reached the highest point of the mountain, there was no sign of the boy at all. He vanished like smoke. Although it was winter, the weather was very hot and sunny. Gasping for air, I felt the warmth of the sun while climbing up the mountain. On one side of the Safi mountains, there was nothing to see except for hundreds of mountain peaks and hills. Far away on the other side was my hometown, the beautiful city of Kabul, where the history of the disaster, the so-called revolution and catastrophe had begun.

What should I do? Where should I go? Back in Kabul, Father, Mother, family, and friends like Hamed, Khawja, Wais, Homayoun, Habib (son of Haji), girlfriends like Latifa, Nasrin, Maryam, Pashtun, Nafiseh, and Nahid. In beautiful Kabul, and home sweet home. Ahead of me was darkness. Nothing else. Breathless with dry lips, haplessly, I looked all around me at the vast mountain ranges. There is no going back. My mind was flooded with thoughts. With what face shall I return home? And where would I go if I don't go back home? At this time, I recalled the scene of my farewell with my grieving family. I saw the tear-filled faces of my father, mother, cousin, and especially my young nephew. Suddenly, I screamed and cried with utter despair. I was sobbing like a child. Without thinking, I jumped to the other side of the mountain. I fell, got up again, and went down. My elbows and legs were bloody. Finally, the tears dried up in my eyes and I reached the end of the mountain. I remembered the words of those two guys who said: "Just follow the trail of the footprints and you will reach the committee." I had no choice but to march ahead. I saw a straight line on the other side of the mountain. The feeling of hunger, tiredness and despair was unbearable. I had no energy left for walking in my legs. The weather was getting cold and the evening was coming closer. The hope of survival was slowly dwindling in me. I was completely disappointed and all my hope was almost dead. Suddenly, I saw two men and a child with two donkeys. Am I hallucinating!? I could not believe my eyes. Indeed, they were farmers, and indeed, they appeared as angels of salvation.

I went to them and greeted them by saying salam. I told my story to them and soon realized that they didn't understand my language. They were Pashtuns and did not understand Farsi, so showing them the letter was useless. From all my conversations, they recognized the Commander, Mullah Mohammad. They pointed out that I should sit down until they load up the packs of shrubs behind the donkey. At the same time, they gave me two pieces of dry bread that were harder than rocks. During these last forty years, I ate the most delicious foods from all around the world, but I will never forget the taste of that dry bread until the end of my life.

We walked for an hour. The sunshine disappeared and on one of those hills a one-story house was seen from the distance. They pointed to that house and repeatedly said. "That's Mullah Mohammed.» They separated from me and went on their own way. The same trail of footprints ended at the back door of that house. I approached the gate. A young man with a long beard grumpily came out of the gate pointing a Kalashnikov at me. He stopped me and asked: «Ta Sok yee?» In Pashto, this means who are you? I said to him, "Please give this letter to the Commander Mr. Mullah Mohammad." He took the letter from my hand and went inside the house. I had no energy left in me to stand on my feet so I collapsed on the ground.

CHAPTER FIVE

I waited about five or ten minutes, sitting on the ground. Finally, the door opened. The same mujahid called me into the house. It was a large house of about three acres of land with two buildings, a small one with two gates, one of which was locked, that later I found out to be a shop, and two small holes as windows close to the roof which were covered with plastic and a hole for a smoke. The other building was large. The gate of the room was higher than the ground, which was connected via stairs. The window of this house was not made of glass; instead, it was covered with plastic too. The restroom was outside the house on the other corner of the yard. In the middle of the yard, between the two houses, there was a huge machine-gun hidden with a special cover, making it unidentifiable by the air force.

Two Mujahideen were taking ablution in the yard. We walked inside the large building. The room was rectangular and inside it, there was a small room within the big room used as a kitchen. The room was heated by a wood-burner. There were approximately fifteen Mujahideen. The men were sitting in a circle with three or more of them busy in the kitchen. Altogether, there were about twenty Mujahideen, all armed with Kalashnikovs.

I put my shoes inside the room near the door. I said salam to them and I received the answer back from everyone. Everyone looked at me. A thin man with a black long beard and white turban had my piece of paper in his hand. He asked me with a Pashto accent, "Are you the relative of Hazrat Sahib?» I answered, "Yes!" Without any introduction, he said, «These days, no one is going to Pakistan because the weather is cold. You have to wait a few days. Then, someone will go.» I wondered how he read my handwriting or if he just understood the word Hazrat Sahib.

Commander Mullah Mohammad was a serious man. He was like a quiet dictator. A young fat child was eating bread with tea in the corner of the room. They told me to sit down and they brought me tea and bread. The child, with a piece of bread in his mouth, said to me, "So that was you shouting to me? I thought you were just saying hello to me. That's why I didn't stop." This was the child who was a foolish guide, the one who made me fall to my knees and cry, the one who made me starve. If I was anywhere else, I would've slapped him with all the power I could. Evening came and dinner was brought to room. It was 'Shorba-e-Landy' soup (dry meat). After a starved day, it tasted phenomenal, whether it truly was or wasn't.

Over the next few days, the meals were the same. Breakfast was sweet tea with bread. Lunch included some variety of dishes. Dinner was always 'Shorba-e-Landy'. I had to be cautious while chewing because the bread and rice had rocks. After eating the night's bread, the time came for prayer. From this moment on, my main problem began. With the crowd, I went out for ablution, but because I didn't know how to do ablution, I looked to my sides and all I knew to do was copy the person next to me. Fortunately, I learned ablution quickly and without others realizing I didn't know how to do so prior. We went back inside the large building; Mullah Mohammad was in front. The rest of us lined up and started the prayers. (congregation pray:

It's the Muslims' communal prayer in a Masjid
whereby they stand in straight rows behind an Imam who leads the

prayer. Muslims are encouraged to pray all of their five daily prayers and obligated to pray the Friday prayer in a mosque in a congregation as it is considered more virtuous than to pray alone.

And the prayers behind imam are silent just follow the imam, not entirely but generally, when the Imam recites, they are silent). There were no electric lights; instead, oil lamps were used. The Mujahideen ranged from seventeen to thirty years old. After their prayers, they would usually gather and tell stories and tales of their battles against government forces or they'd play games suitable for children. A favorite of everyone was a guessing game that involved four men on one side and four on the other side that were hiding something in their hands under a cover. One group of four had to guess what was in the hands of the other four. The loser had to be slapped behind the hand.

I slept in that small building. The room was heated with a stove. During this cold winter, the warmth was suitable in the room. I did not feel cold at all all night. Most of the quilts were Pakistani made. They gave me a mattress and a pillow. I always slept with my bag of personal belongings under the pillow.

CHAPTER SIX

The next morning, everyone woke up for prayer. I was following the people's movements. After morning prayer, a number of them slept again while others went out. I didn't know where they were going. A number of men were busy with household chores. Those were all Afghans, but their characteristics and behavior did not resemble my family, friends, nor the community I grew up in. They did not speak much. They did not know the joy of joking and laughter that I was accustomed to. They did not have the slightest interest in music. Most importantly, there was a complete lack of existence of any woman in their life. These men did not know anything about women. I came to the conclusion that they were born from the trees or the stones or somehow the birth of these men was a divine miracle that didn't come through a woman. The local store was with limited choices for shopping, including only cakes, cigarettes, and a few common necessities. Although I had few packs of cigarettes with me, the existence of the cigarettes in that shop made me feel more at ease. One Mujahid was from Herat. We built a bit of friendship. If any day Mullah Mohammad was away from the house, we would come out to the yard and he would do target shooting with a gun. Sometimes he even allowed me to shoot a couple of rounds. This was the first time in my life that I handled a gun in my life.

From the distance and from the top of the hills I could see that nearly every few miles there were mud houses on the hills. One day, I asked my new friend who lived in those. He answered, «Every house belongs to a different group of freedom fighters. The only one that has no house in the area is that of the Islamic Party, a fierce rival party." He spoke about a fight between the Mujahideen and the Islamic Party of Gulbudin Hekmatyar where after two nights of fighting, the Mujahideen managed to defeat them into forced retreat. He continued, "You were fortunate not to have come across them along the way. If you were captured with the letter of Jabha-e-Milli Nejat at your possession, the Islamic party would have killed you."

During my time there, I often sat next to the wall sunbathing, thinking about Kabul. My mother, my dad, my friends, the new city, and my room were the center of my world. I was wondering about my friends and what they were doing at this moment, and whether the girls I knew from the girls' schools Zarghona and Malalai think of me in my absence. I wondered if our home phone number, 31639, was still active and ringing since I left. Thousands of sweet thoughts would come to me and I would hide my tears from others.

I was consumed by memories of the last few months of my time in Kabul. One of such was how my family had become friends with Zarmina and Amena, the daughters of Sayed Hakim, (The owner of the Hakim Dry-Cleaning). Sayed Hakim's laundromat was on the next street from where we lived. Our friendship was forged from the common interest of fleeing Afghanistan during these violent and turbulent times. My sister and I become friends with Zarmina very quickly. Zarmina often liked to stay in our house. Zarmina had very sweet and lovely voice and Khwaja Hamid also had a talent for singing. Our house was always filled with music, my grandmother, my mother's mother, had a cheerful soul. She liked Khwaja's voice and if anyone else tried to sing, she would say, "Your kiddos be silent! You be lucky to sing like him.»

I wondered how circumstances can change humans. I was a naughty and cowardly child who feared everything: darkness, thieves, and the horrors of ghosts would haunt my mind so much that instead of going outside to use the toilet, I would just urinate out of my window onto the neighbor's wall. But now, far away from home, facing massive uncertainty and the accepting thousands of dangers, I was heading further out to an unknown future towards a foreign country.

CHAPTER SEVEN

I've spent many days in the world of my imagination. My prayer skills were improving. It's been two weeks now that I've been waiting without any mention of a chance to go to Pakistan. My family had no news of me, whether I was dead or alive. Although it was the month of January, the weather was mostly sunny and bright. Only once or twice, the air became cloudy and it snowed, but snow did not pile on the ground.

One sunny day, everyone was busy with their usual chores. Suddenly, two Russian helicopters hovered in the sky above the house. They turned in a specific area between the hills. Mujahideen and housemates rushed in a panic, sprawling around like ants running crazily. After about fifteen minutes, the helicopters and the sounds of their rotors disappeared. A couple of men hurriedly attempted to conceal the machine gun even more than it already was. A few minutes after the helicopters left, the peaceful silence was suddenly disturbed by the loud, ferocious roaring of two jet aircrafts. Like lightning, they appeared with loud sounds bombarding the plains between the two hills. The Mujahideen scurried in all directions to find a place to hide. Until that moment, I did not know that there was a hole in the yard. They rushed into that hole. I also wanted to go with them, but wouldn't let mee. Instead, they left me behind with the shopkeeper. Perhaps that cave was their secret stash of weapons. God knows what else was down there. They ordered us to walk into the courtyard. The aircrafts left the area, leaving behind the harsh sounds of their bombs. As the planes went off, the Mujahideen came out of the hole. I asked my Herati friend who and what was up that hill. He replied, "There is 'Jamaat Islami Committee', another Mujahideen group." One or two hours later, I found out that only one sheep was dead and there were no other casualties.

It's been three weeks with the Mujahideen. One night before sunset,

a group of over twenty Mujahideen arrived in the house. I asked my Herati friend, "Who are they?" He said, "They have returned from jihad and some days from now, another group will go in their place." With their arrival, everything changed from dinner to sleeping arrangements. I had suspicions that many of them were homosexual. Because they were too many of them in one shared room to sleep, they all slept close together. I became hyper vigilant and scared. I had trouble sleeping and slept with my back to the floor to save my dignity. Frankly, with the arrival of these wild looking men, I dreaded nights.

I was a sorry sight. My clothes were so dirty that the color of my clothes turned greyish-black. My beard grew longer and fortunately, I have had a very thick beard which covered my face, a sure safety net from the prying eyes. After a while, the commander Mullah Mohammad called me into the room. I got in and I saw three new faces. He told me that in a few days, the commander of the unit, Commander Wahed, who was one of these three men, would take me to Pakistan with him. I was surprised and the same time worried because I was feeling safe here. I didn't know them and if I could trust them. At the same time, I was eager to move on.

One day, I was surprised to see merriment fill every corner of the room of that bleak house. I asked someone about the cause of all this joy. They responded, "Mullah Mohammad will get married for the second time tonight!" Some men were selected to go with Mullah Mohammad to Dehsabz for the wedding. They decided they'd get moving after the evening prayers. As the evening prayer began, I copied them, just like every time, and I thought that my copying was really good after many weeks of praying. But I was shocked to find myself addressed by one of the three Mujahid's after the end of the prayer. He shouted at me with an angry, loud voice that everyone heard. «How do you pray? Do you know the prayer or not?» I did not know what was wrong about the way I prayed that he had noticed with such distinction. Everyone looked at me as I stood in total cluelessness, paralyzed in my response. Mullah Mohammed came to my rescue and said, «It is none of your business, anyone can offer the prayers as he wishes!». The Mujahid didn't make any further comments and I felt a huge sense of relief overcome me.

The wedding group left the house for the Dehsabz. We slept for one more night at Mullah's house. The next day during the afternoon, Commander Wahed, two other men, and I set off for Pakistan.

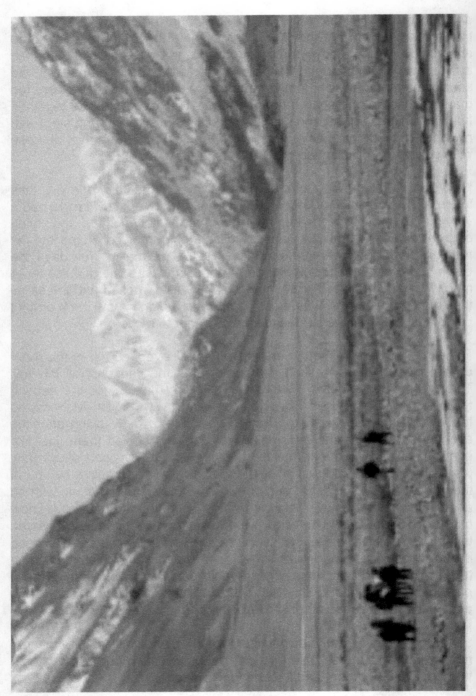

CHAPTER EIGHT

We around 12 o'clock, perhaps earlier than that left the Jabha-e- Milli Nijat committee. After an hour's walk later, we arrived at another committee and entered a house that was built in the middle of the mountain. The Mujahedeen there, who turned out to be older men, had just prepared lunch. They were speaking in Pashto, sometimes about me. After the lunch and few cups of tea, prayer was called. After that, we said goodbye to them and left.

Commander Wahed was a very tall man with wide shoulders, calm, mysterious, and sensational. The Commander went ahead and we walked after him. The two other companions sometimes talked among themselves. Sometimes they'd ask how I was doing. Commander Wahed was often silent.

Gradually, we were making our way down from the mountains to the ground. The sky was bright and blue. The air was fairly cold. As I followed the men down the mountain, I was lost in the beauty of nature, like I was living in a fantasy world. The beautiful landscapes snatched the sorrow of loneliness from me.

At a hillside of the mountain near the Mahipar (Mahipar Dam It is located 40km away from Kabul. This dam was built in 1344 in cooperation with Germany and has 3 turbines with a power of 66 megawatts of electricity. Currently, a 19-megawatt turbine is active, the rest is worn out. This dam does not have a natural water storage. It is extracted from the water pump by the water.) There was an immense and beautiful garden with a tranquil atmosphere. Commander Wahed told me that this garden belonged to King Zahir Shah. There was no road to the garden. I thought maybe the king would come with a helicopter. We walked on a seemingly endless path. I was so tired that my five-kilogram bag felt like an immense burden. Commander Wahed noticed my predicament and ordered one of the men to take my bag. From that moment on until we reached to Pakistan, the bag was taking turns being carried by the two men that accompanied the Commander.

The sun slowly crept behind the mountains and the air suddenly got colder as nightfall approached. In the late hours of the night, a group of five people joined us. They talked to each other in Pashto and mentioned

where they were heading. About half-an-hour later, we were halted by Commander Wahed. He told us that from now on, there shouldn't be any fires or noises. "When we get to the bridge, we'll run across as fast as we can. Once on the other side, we must stay away from the road and head deeper into the agricultural lands as far as possible. At all costs, we must avoid the road to be safe from the government checkpoints and the soldiers. We walked for about twenty minutes to get near the bridge. Once there, we dispersed and tried to hide behind bushes. It was very dark so it was difficult to see what was in front of us. We were whispering in each other's ears to help one another. Then it came time to cross the bridge.

The bridge was made of wood and ropes. It was not made for easy crossing. Beforehand, on our walk over here, I was told many tales about fellow citizens who died in this same spot. We had to run one by one across the bridge, except they didn't let me cross it alone. One of the two Mujahid was assigned to be with me. When our turn came, we sprinted to the bridge. On the bridge, my hand was in the Mujahid's hand as we ran on the bridge. One of my feet got stuck in hole. The Mujahid mumbled slowly, «What are you doing?» I said «Nothing. It's not my fault my foot is stuck in a hole." Fortunately, the sound of the gushing water underneath was so loud that the government's troops could not hear our voices.

He helped me pull my foot out of the hole. My foot was slightly scratched. As we got ourselves to the other side of the road, he left my hand. We came by the fields and without looking back, we started sprinting again. The agricultural land was a terraced field with patches that followed each other like giant steps. The muddy and rippling ground was painful to my feet. We ran fast in the darkness and each time with the difference in the surface of the patches.

Every now and then, I'd get stuck or I'd fall down on my face. Each time, the mujahid would say in anger, «O' kiddo, you're making everyone aware of us!» Eventually, after we ran non-stop for a few miles, we rejoined with those who were ahead of us. The Mujahid was telling others how I fell many times and they snickered quietly. We spent a few more hours in the dark until the Commander allowed us to light a cigarette again. I don't remember when we were separated from the group of people who joined us on the way, but they weren't with us anymore.

The morning air was piercing cold and I was not wearing winter clothes.

My body felt the cold air down to my bone marrow. My face was completely frozen. My hands were completely numb. I smoked cigarettes constantly so that I could get a temporary dim feeling of warmth on my face. I opened the jacket chains and hid my hands under my armpits. We walked all day till the late evening. The night was closing in and the air was sharpening with coldness. Not only my eyes, but all the cells of my body craved for sleep. The only reason I was still able to walk was out of fear of what'd happen if I stopped and the hope of getting somewhere safe.

We carried on walking. The sky was clear and it seemed as if stars are celebrating. That night, we walked in silence sometimes beside each other and sometimes after one another. Suddenly, we heard the sound of a groan. Commander Wahed and others talked to each other in Pashto. I could not comprehend a word of what they were saying, but their movements were animated. They were surprised too. The sound of the groan was getting closer and closer as we walked. In that absolutely quiet overshoot, the groans reflected a very sad, annoying, and horrific sound. All four of us were curious to know who was in pain and why they were groaning. Now, the sound of the steps was heard. That's when a group of about fifteen people from Northern Afghanistan emerged. They had a horse and a donkey with them. At the back of a horse, a coffin was tied up. The same of the groans was raised from the coffin. Commander Wahed spoke to them, asking what was the matter for the coffin. They told us that they were from the North and that a young man from their family had stepped on the mine, causing all of his body to be injured by the explosion. They mentioned they were taking him to Pakistan for treatment.

I saw that poor young man who looked like he was between twenty to twenty-three years old. They were carrying a coffin full of cotton in which they placed the wounded young man on top of. I thought they were carrying him in the coffin because if he didn't make it alive to the treatment facility, they could bury him wherever possible. Because of the weather and the cold wind, his wounds were agonizing him even more. His groans were getting so loud that they could be heard as high up as the seven heavens. Seeing and hearing this scene made me feel so terrible. Due to the war between the two parties and the fear of the Islamic Party, this family took a long detour from Northern Afghanistan to reach Pakistan. Since our destination was the same, we decided to travel together.

Faraway in the darkness of the night, a dim light could be seen. The

Commander said something to his friends and then he addressed me. "There is an abandoned school. It seems people had started a fire there. We are going there to get some rest." After nearly fifteen hours of walking with little rest, this is all I wanted. The group with the injured man refused to rest and decided to continue their journey to Pakistan. We said our farewells and wished each other a safe voyage.

The school was punctured by thousands of bullet holes. There were no doors, no windows, and only some rooms had a roof. In each room, people started a fire and sat around in groups of five to ten people. I don't know if it was the mind-altering effects of walking for hours sleep-deprived or the blistering cold, and in every room, we entered, the people who were around the fire resembled zombies to me. All of them looked at me and I felt very scared. We went from one room to another and after circulating all the rooms, the commander asked me if I wanted to stay here. I promptly told him that I'm not tired. The truth was I felt like I was dying from exhaustion. The commander seemed to understand exactly how I really felt. We got out of that dreaded place and returned back to pitch black silence.

After a couple of hours walking, I started struggling to keep my eyes open. I felt choked by the smoke of my cigarette. I tucked my hands under the refuge of my armpits again. My eyes were heavy and finally, I fell asleep while walking until all of a sudden, the ice broke under my feet and I fell down into the freezing water. The water was so cold that my left foot got frozen through my sock and shoe. As my eyes opened, my right foot fell into the water too. Both my feet froze instantly. When the sleep overcame me, the cigarette fell from my mouth. I thought I was drawing my last breath.

At that moment, I heard the voice of Commander Wahed saying, «We have reached!» I did not know what he meant. Where did we get to? He said, "Can't you see we have reached the village of Tizin?" I didn't even know what time in the morning it was. As much as I rubbed my eyes to see something, I could not see, as they were blinded by darkness. We walked further for five miles before I could identify a fairly large village surrounded by trees and snow. From one or two of the roofs' chimneys, I saw smoke coming out and it made me feel warm inside. I didn't have the energy to walk and I couldn't feel my feet due to the harsh weather. I was eager to take off my shoes at my earliest opportunity.

We got inside a mosque and we were greeted warmly. Commander Wahed told the Mullah who he was and where he was going. The Mullah told us that we have arrived at a good time and asked us to go and get ablution because it was time for morning prayers. We left the mosque to take ablutions. A small stream was flowing near the mosque. For ablution you have to break the ice. After ablution in that cold water, I had no other organ left that was not frozen. I thought to myself that if the Prophet of Islam were here instead of Saudi Arabia, there would be a verse in Qur'an that ablution would be forbidden by the water underneath the ice because everything below the bellybutton and upper the knee would get frozen.

CHAPTER NINE

The temperature inside the mosque was comfortable and warm. The mosque had the capacity for approximately thirty worshippers. The mosque was traditionally built higher up, at its cellar that fire was burnt, which warmed the whole mosque at once. In Kabul, this type of underfloor heating is called «tawa khana».

After taking off my shoes, my feet and legs felt extremely relieved after all the walking. I put my shoes and socks near the fire to dry. My pants were also wet, but they were not so bothersome. After the morning prayers, they gave us milky tea and bread. Then, they left us to catch up with some sleep. The noise inside and outside the mosque quietly calmed down; only occasionally children's' voices were heard. I was laying in a corner like a dead piece of meat and I fell asleep within a second.

People's voices woke me up. I saw that the Commander and the other two Mujahids had just been awake and sitting in their places. Men were preparing for the afternoon prayer. We went out for refreshments and ablution. The air outside was bone-chillingly cold. The congregation prayed. After praying, they gave us lunch. My shoes and socks were slightly dry. I really wanted to continue with our journey after lunch. I asked the commander, "Should I wear my shoes and get ready to go?" He said, «We'll stay here today and we'll leave tomorrow.» In a place of safety and refuge, I couldn't understand why I felt so nervous. I didn't like staying here anymore, but the decision was in his hands. We slept a little after the lunch.

For the first time, Commander Wahed was in a mood for conversation. After the afternoon nap, he asked me, «How was your last night journey destination?» I replied, «I thought that we'd walk for ten or twelve hours and after getting some rest, we'd begin the journey again. I didn't know that we'd travel non-stop for sixteen hours all night, through pitch-black darkness and such cold weather." The commander said, "I saw that sleep was overcoming you in the morning and you were very tired. Why were you so keen to leave the school? Didn't you want rest?" I told him that the people there were so scary. The Commander and others laughed. One of the Mujahid asked me, "Don't you feel cold in the head without a hat?" The other Mujahid replied, "He is like the Hindus. They keep their feet warm and don't cover their head. Don't you see he's wearing two pairs of socks?" I just smiled at them.

Everyone gathered for evening prayer. We went outside and broke the ice to take an ablution. After completing the prayer, a boy of twelve years old beside me noticed the fault in my prayers and revealed this great mystery to all. A meeting began in the mosque. I went outside with anger and fear from the mosque and sit in a sunlit balcony. Their loud and animated voices discussing like members of an important meeting could be heard. They were talking in Pashto, and even though I didn't speak Pashto, I almost knew what they were saying. Mullah's proposal for Commander Wahed was to leave me with them and for the Commander to go to Pakistan to finish his work and come back for me and take me with him. Mullah said that we couldn't trust this guy, referring to me. I heard him ask the Commander, "We don't know what he is? What's his religion? Is he a spy? Is he after discovering secrets of the way of Muslim brothers journeying to Pakistan?"

Everyone had a comment to share at that meeting. I was so afraid of listening to them. Suddenly, the same kid came out. He looked at me as if I was the enemy of God and Prophet. He asked me in strong Pashto accent, "You don't know the prayers? I noticed you this afternoon that you don't sit correctly and your forehead doesn't touch the floor." Then I realized that the time of sajdah, I didn't put my head properly on the floor, and this little boy, without concern for his own prayer, noticed me. No one asked him why he was not attentive to his God while praying, and why he'd rather focus on others.

Commander Wahed assured them that I am not a spy because I belong to

Hazrat's family and Commander Mullah Mohammad, who is a Muslim and a Mujahid from Dehsabz, handed him to me. The Commander continued saying, «If he were a spy, in any case, Mullah Mohammad would never send him with us. He is going to Hazrat Sahib in Pakistan." I did not dare to enter the mosque until dark.

Finally, the temperature cooled down and I was forced to go inside the mosque. My head was lowered and I stared at the floor the entire time. They brought the dinner. Then, the prayer was read. This time, I was very cautious with my copying of others and I didn't make mistakes. I pressed my forehead to the ground while my brain was saying «Subhan Allah». We slept in the mosque that night. The next day after morning prayers, we resumed our journey. Two hours later, we reached the foot of the famous mountain 'Safid Koh', the White Mountain. I heard that the mountain is always covered with snow and that's why it was called «White Mountain». One of the two Mujahids told me that in this mountain, many animals and Mujahideen had lost their lives because of treacherous conditions. "If one foot slipped or got stuck in the snow, the person's salvation would be in God's hand," said the Mujahid. I was surprised to find that in the middle of the mountain, a single restaurant stood with open doors. We were served kebab for lunch. After lunch, we continued our trek. On one hand, I was worried that we may spend another night in a mosque, and on the other hand, I was scared of traveling all night.

Despite nightfall and the chilling weather making the walk difficult, we continued on without interruption. It was around ten in the evening. The cold crept up on us on one side and exhaustion on the other. It felt like the two wolves eating us from inside our bones. As we were walking, a light appeared from the holes of a small hut. We had to cross a small ravine over which a single plank wood was laid precariously which purposed as a bridge. The size of the plank was as big as a large tree. If it was not winter, it wouldn't have been as difficult to cross it easily, but because it was winter, it was very difficult to pass through. At ten o'clock, in the dark of winter, the river water made a slippery ice pipe that was more dangerous than the Puli Salat (some imaginary bridge in the hell that sinners must pass according to Islamic beliefs). It didn't seem like it was meant to be a mean feat. It was about six feet long. Everyone passed through it, except me. I slipped down with one foot in the water. I was hit hard by the wooden plank and got hurt in my sensitive parts. A thought struck me that because of this hit, I may never have children.

Thankfully, it wasn't just me who slipped. After me, one of the Mujahid's foot was also in water. I felt a sense of relief that I wasn't the only one to be laughed at. We got inside the cottage where we saw four other men lounging inside. I don't remember where this cottage was and why it was built. It was just a single room that had a Jute carpet and a wooden stove that was burning in a corner to provide warmth in the hut. I put my shoes close to the heater. Everyone immediately fell asleep wherever we sat. Honestly, we didn't sleep, but in the middle of that chilly and scary mountain between sleep and waking we made it through to the morning.

CHAPTER TEN

We carried on our journey like always, just after the morning prayer. The temperature gradually became warmer and more pleasant. There was no white snow, but the mountains were covered by green shrubs and mulberry trees. We were passing through the Paktia province and Jaji Mangal area. The weather was gorgeous and the sights and scenery were breathtakingly picturesque and beautiful. Everyone suddenly got jolly and talkative. Humans are awesome creatures. As they see beauty, they forget about everything else. All problems seem to disappear.

The weather was so good and warm that I had to take off my cotton-padded coat and forcefully push it into my small bag. At noon, we arrived to a burnt-up school with no teachers nor students. Our company had some pieces of bread and a few boiled eggs. We ate our lunch and then moved on. In the evening, at about five o'clock, we arrived at an area called Parachinar. Finally, after about one month being on this journey, I saw a city and a community of people. We crossed a line, which the commander mentioned was the border of Afghanistan and Pakistan. The Afghan side had no check posts nor any guards. The Pakistani side had three officers who weren't wearing uniforms. It was difficult to distinguish them from ordinary people. They were wearing a dirty shirt, a green knitted jacket worn over their shoulders, a waistcoat, and black "chapal" (sandals), They reminded me of the comedians from the Hindi films. You would've thought the Pakistani government might have told them to guard the border, wearing a uniform suitable for the task.

Men, children, and women in nomadic dresses were passing the border

without any concern. There were «food» shops in Pakistan's side, one or two weapon and arms shops, and several restaurants. We went to the second floor of one of these restaurants. The restaurant was almost full. After a few weeks, I heard music for the first time, which was played from a torn loudspeaker. The sound of Peshawari music was very annoying, but I noticed some people were listening to it and enjoying it. I said to myself, I'll tolerate this and hope this toleration may not be for long.

Commander Wahed and his men seemed to be very happy by their reactions to the taste of the food. They spoke about how they were looking forward to having a meal here throughout the entire journey. They ate the food with such a voracious appetite until there wasn't even a scrap of food left. After drinking tea and paying, Commander Wahed paid for me too. We went to Hadda of Peshawar (bus stations are called Hadda). The van driver had already filled the van with passengers, but somehow accommodated us four too. In the gap between the seats and walkways, they made extra makeshift seats from wooden boards that didn't have a back and were not fixed properly. The unfortunate passengers sitting on these seats had to sit up-right for four hours through rough unpaved roads. The way Pakistani drivers drive is treacherous. After reaching the destination, one would be totally numb, unable to have any feeling left in his back or his bottom for a while.

We arrived in Peshawar late in the evening. The city was crowded with people moving in all directions. It was apparent that the city's population was more than its capacity. We rented two rooms in a shabby hotel to sleep for the night. The voice of the music and the Quranic recitations were heard all night long from the loudspeakers. I quickly fell asleep.

I felt that someone was shaking me. I opened my eyes. Commander Wahed woke me up to prepare for morning prayer. I went to the bathroom. I didn't know how much time it took when I returned, but I saw that the commander and his friends were not there. They appeared after some time. The Commander asked with anger, "Why didn't you offer the prayers with us?" I told him my reason and convinced Commander Wahed. He said, "I don't know but God knows well what your religion is. Get ready. We're going to the office of 'Ethade Se-gana' (Union of three Mujahideen Parties) to see Hazrat Sahib."

CHAPTER ELEVEN

We took a rickshaw to the 'Etehade Segana'. The City of Peshawar is the city of business, rushing, noisiness, and stray dogs, although unlike their police, the dogs didn't annoy anyone. Sometimes you could see someone pulling a goat with a rope around its neck. At first glance, the goat looked like a dog. There was trash in every corner and nobody paid any attention to it. People were so accustomed to it that trash was a part Peshawar's identity as a city. All the dirty and low-level jobs were fulfilled by Christians, such as cleaning the streets, taking the trash, and even cleaning latrines. Some of Pakistani toilets were not built like the ones in Afghanistan. In Afghanistan, the toilets had a hole for keeping human waste in a separate section. In Pakistan, they would build a little open-top pedestal off the ground using bricks. The waste would be collected right there and Christian workers were assigned with cleaning it.

The sounds of the tragic music was heard from restaurants and cars alike and the Quranic recitation was heard from loud speakers broadcasted from mosque minarets without interruption. The smoke from burning wood, plastic, paper, leaves and twigs was everywhere. It was mosquito season and so peopled burned anything they had to repel mosquitoes.

I noticed uniformity was another characteristic of Peshawar. One would think that all the men in the city were in the same shape, dress, morals, and customs alike. Among the two hundred men, only one or two women would be seen, most of whom were Afghan refugee women or Christian women doing their chores. This place was a jungle of its own and it was necessary to learn the ways of survival in this jungle. I was drowned in my thoughts until I realized we reached the 'Etehad Segana' Building.

This was a very large building with tall walls. The guards drew my attention. It consisted of two houses having two stories with several rooms used as offices. The rooms and corridors were swarmed with hundreds of men. Many of them were moving from one office to another. Later, I found out that they were in a hurry to get the weapon in the name of Jihad, either to sell it at the frontier areas or destroy a village, town, city or region. Everyone spokes in Pashto. Occasionally

Me in Peshawar city at the age of seventeen

some men smiled. Otherwise, the rest were very serious, sombre, tense, and even scary.

We passed through the crowd and entered the yard. We went up to the second floor and entered into an office with closed doors. Commander Wahed went to the person behind a desk and pointed at me. "This person belongs to Hazrat Sahib's family and he wants to see Hazart Sahib Zabihullah, the son of the Hazrat Sahib." This was Hazrat Zabihullah's office. He asked if I knew him personally. I told him no. After a while, someone took me to Hazrat Zabihullah's office.

It was a regular office with a desk, several chairs, and a couch arranged around the desk that was for guests. On the other side, a young man was sitting behind a large table. Hazrat Zabihullah was around 30 years of age. He was thin, bearded, and had a hat of Pakol. I greeted him and he rose halfway to give me a handshake before pointing to have a seat. He introduced himself as Hazrat Zabihullah. He asked me, "Are you the one who knows my aunts and grandmother?" I answered back with a yes. I began to tell the story of the friendship between our families with his aunt Asma and Atiqa his grandmother. Zabihullah listened politely to my words, then he said, «I don't know about the relationship between our families in Kabul. You should go to Islamabad to our house.» Zabi addressed a young man in the room and told him, "Send Khalil to Islamabad and tell the bus driver to cooperate with him in order to reach his destination". After this he stood up and gave me a handshake for a farewell.

Commander Wahed[3] was waiting for me with his companions outside the office. The young man, who may have been the assistant of Zabihullah Mojaddedi, told them what Zabihullah's instruction were. The Commander and his companions gave me a very warm hug and we parted ways.

We took a rickshaw to Islamabad's Hadda where we boarded a bus with a GTS (Government Transport Service) logo. My companion said something to the driver and then he paid for my travel fare. He told me that this bus is going to Rawalpindi. "The driver will drop you where you need to take another bus to Islamabad, and then he'll instruct you where to get off in front of Hazrat Sahib house. I've paid the bus fare for Rawalpindi to Islamabad to the driver." He gave me the address of the house on a piece of paper.

[3] In that moment, I knew I would never forget the help and affection of Commander Wahed. It's because of him that I am still alive. Long after I heard that Commander Wahed was killed in the battles. If it's true, may his soul rest in peace, and if it's not true, wherever he is, he may live a healthy, happy and peaceful life.

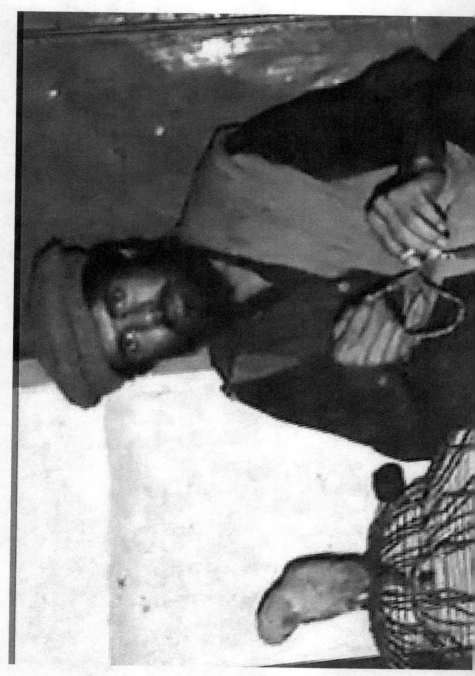

Me in Peshawar city at the age of seventeen

CHAPTER TWELVE

The bus was full after several hours and moved to Rawalpindi. We didn't get away for even a few minutes from Peshawar before I felt an urgent and intense need to go to washroom immediately. With every shake from the rough unpaved roads, the risk of wetting myself was increasing rapidly. Gradually, I felt the pain between my legs become unbearable to a point that my body was twisting like a snake. The Punjabi thin man who was sitting besides me noticed my pain, looked at me, and told me something in Punjabi. I didn't know how to tell the driver that I needed to go to washroom very badly. Most Afghans, including myself, believe that by watching all those Indian movies, we could speak Urdu fluently. It was partly true, since I had just heard it but never spoke it, so I could understand half of Urdu but couldn't speak it.

Throughout the whole ride, I was wondering what the word for urinating in Urdu was. While I was contorted in pain, I was thinking of Amitabh Bachchan and Dharmendra (two famous Bollywood actors) to recall if they ever mentioned the word for urinating in their movies, but I couldn't remember. At that moment of my suffering, from the bottom of my heart as cursing all the Indian actors, their sisters and their mothers for not ever mentioning the word for urinating even once. I thought of going and telling driver that I want to piss several times, but I figured he wouldn't understand me. While fellow passengers were enjoying their travel, I was suffering from my pain.

After an hour, we arrived at the first city, Faizabad. The Peshawari buses were taking a rest stop here, so I quickly rushed to the 'washrooms' in lightning speed. I became calm and relaxed. It was there that I understood the Poet Saadi's words: "If a part of the body is sore, the rest of the organs are feeling the pain too." I came back to my seat and the bus drove onwards[4]. That is a part of my memories of life which is tied to the piss.

Now, feeling some comfort finally, I looked out from the window of the bus to the sights and noticed that on all the walls alongside the road, things were written in Urdu in black small and large fonts. The most frequent messages were "Yaha Pishab Karna Mana Hay" and «Mardana Kamzori ka Elaj". I guessed that these phrases may signify the names

---

[4] It has been thirty years since that day but I have not forgotten and still can feel that pain.

of the owners of those houses, but a few months later I got to know the meaning of them. The first phrase meant "You're not allowed to pee here" and the second phrase meant "Treatment of Men's Sexual Weakness».

We reached the city of Rawalpindi at Peshawar Hadda. From there, the driver of the bus walked me to another bus and told the bus conductor something I couldn't understand and put something in his hand that I couldn't see. The bus moved and I had fallen into a deep silence and felt very lonely. I wanted to scream and cry. I didn't know where I was going nor what I was doing. I thought to myself, if I don't find Sibghatullah' sisters until night, where will I go and what will I do? I was drowning in anxiety and then suddenly, out of the bus window, I saw Mohammed Jan Goren, the well-known actor of Afghan Radio and Theatre. He was a friend and theatre colleague of my elder brother. He was buying fruit. Reaction took over me and I got out of the bus and reached for him at lightning speed, grabbing him like a child who had lost his mother and found her again.

With tears in my eyes, I took him in my arms and didn't want to let him go. Mohammed Jan Goren didn't recognize me in this strange place. He was surprised and shocked when he saw me. After a minute or two, I calmed down. Then, attempting to maintain my composure as best as possible, I introduced myself. He asked about my brother, whether he was released from prison, and asked me where I was going. I told him my story. He told me that I shouldn't have left that bus because it would've taken me to the correct address. He was accompanied by another young man. They walked me back to the bus stop. When the bus came, Mohammed Jan Goren talked with the driver in Urdu and gave him some money. He told me not to step off the bus until the bus driver instructs me to.

Although I didn't want to be separated from them, I had to travel to my destination. The bus finally arrived from Rawalpindi to Islamabad. The driver told me something in Urdu and I realized from his gestures that I had to leave now. The bus conductor showed me the road to walk down. I searched and found the house number on that road.

I came across a beautiful two-story house with high walls. Two guards were guarding the house outside the gate. I greeted and gave Hazrat Zabihullah's letter to them. One of them went inside to inform Asma and Atiqa, the sisters of Hazrat Mojaddedi. I took a deep breath. At that moment, I felt calm and assured that my problems and dangers were over. I was lost in my happy thoughts, imagining that Asma and Atiqa came with their glowing faces and hugged me kindly, asking me about my sister and especially my mother, who treated them as daughters. I was taken to the house where all the comforts were awaiting me. I snapped out of my daydream when I suddenly heard the sound of "asSalamualaikum" from a young man.

The young man guided me into the backyard. The house was very similar to the houses of our own in the Wazir Akbar Khan neighborhood of Kabul. On one side, there was a two-story building with windows covered with metal fences. On the left side of the property, there were guest quarters. The two-story building had a garage on the first floor and rooms on the second floor. The young man was walking fast ahead of me and I chased him down. The young man came from Kabul. If he was not Kabuli, he must have lived in Kabul for a long time. We went upstairs and he beckoned me into a room. The room was facing the sun and was covered with a large carpet. In each of the room's four corners, mattresses were laid. It was an en-suite guest room, which had a bathroom with a toilet and a sink. The two windows of the room faced opposite directions with one facing the yard and the other facing the road, with a window on the side of the dining room with a curtain rail, but there were no curtains on the road side window.

There was another guard inside the room. We greeted each other. Three men who looked like constables guarded the house. I took off my shoes and then the introductory questions began. "How are you? How was the journey from Kabul to here?" He then said, "You must be tired. Let me bring hot water and change your clothes." Then, I noticed that the color of garments was dark grey from dirt. It didn't appear to be like cloths made out of ordinary material. It was more resemblant of leather.

The young man was preparing hot water. I felt like taking shower after a long time of having gone without one. After the shower, I felt as if I had lost one-hundred kilograms of weight. I was so exhausted that after I got dressed, I couldn't recall whether I had an extra change of clothes or if they provided me something to wear. I was still eagerly expecting to see both sisters and the mother of Hazrat Sahib. The Young man told me

that Bibi[5] Asma and Bibi Atiqa. couldn't see me and that I'd have to wait another couple of days when «Hazard Sahib Habib» returns from his studies to help me. I had no choice but to wait.

I was in that room with the constables for a couple of days. We ate at the same place and slept in the same room. There was no interest in me from the Hazrat's family. No one came to get me nor was there any direct news from those who professed to be akin to members of my own family. The room was like a prison. From dawn to dusk, morning to evening, I was looking out of the window fence because I wasn't allowed to look at the yard. I wasn't allowed to go out or sit outside the gate. If I were allowed to go out, I didn't know where I would go anyway. I didn't know anything about this foreign land nor anywhere to go.

The next day, while I was in the room, I heard noises from the outside. The young man came to me in a hurry and ordered me not to come close to window because the great Hazrat Sahib may see me. I thought, so what if he sees me. The cars were all lined up next to each other. The guards were armed on each side. For the first time, I heard the voice of Sibghatullah Mojaddedi, who spoke to someone loudly while walking towards the house. A few hours later, they left the area with the same pomp.

The following afternoon, I saw a very beautiful vehicle in red and white color. I didn't know whether it was a Cadillac or Chevrolet. One of the constables said that Hazrat Sahib Habib Agha has come from his studies. One or two hours later, a young fair-skinned guy of about twenty-one years old who had a regularly trimmed beard with long hair and model-like face came into the room and shook my hand. He sat in front of me and with a tone that I couldn't make whether it was friendly, authoritative, or rude, said to me, «You are Khalil. You know my mother from Kabul. My sister Asma says that your family is in Iran. If you want to go abroad, it's easier from Iran. It's not possible to stay here for long. I will send you to the city of Quetta to our committee and I'll write a letter to the committee representative to assist you to go to Iran. He is a good man. Until travel arrangements to Iran are made, you can stay there in the committee. Hazrat Sahib (referring to his elder brother) doesn't like

---

[5] Bibi is an expression of respect, purity and honor generally reserved for pious woman. That's why they'd call an elderly woman who was deserving of all these attributes as Bibi such as grandmothers.

strangers staying in his house."

My heart sunk and all my hope that Hazrat's mother and sister would treat me like a family fell apart. Hazrat Sahib himself is a stranger in Pakistan, but he doesn't want me to stay in his house. I was drowned in my strange and uncomfortable thoughts. Hazrat Sahib Habib Agha was busy telling his story to the constables. He spoke about his education and his place of study. I think he was "studying "in Multan. Then, he spoke in a demanding voice to the young man, "Send Khalil to Hadda Quetta Rail and help him get into the right train." Then, he got up and gave me a reluctant handshake and went out of the room with that young man. Ten minutes later, the young servant came and told me that we had to move to Rawalpindi immediately. I gathered my clothes and accompanied the young man from the palace of Hazrat Sahib, who didn't like «strangers» there, and left for Rawalpindi.

CHAPTER THIRTEEN

We arrived at Rawalpindi train station late in the evening. The young man converted my four-thousand-and-some-hundred Afghanis into Pakistani rupees, which was one-thousand-and-some-hundred rupees. The train was planned to depart the following morning. In front of the train station, there were several hotels. He rented a room for me. The room was very small and had no windows. There was only enough room for a single bed. After being released from Sibghatullah Mojaddedi's "prison", I wanted to go outside and breathe the open air. In the middle of a crowd of people, a young man came and greeted me. I reciprocated his greeting with kindness. I imagined this kind young man to be a friendly individual among the thousands of Pakistanis and that he recognized me as an Afghan. Much to my delight, he was from the neighborhood of "Shar e naw" in Kabul. I told him my story in a matter of minutes and explained why I was staying at this hotel. He advised me to tell the owner of the hotel to wake me up tomorrow morning. He said, "Otherwise, you will oversleep and miss your train." At last, he made sure that I was taken care of and he spoke with the hotel manager to wake me up in the morning. I asked him what he was doing here. "I'm waiting for the US case," he replied. I didn't know what the meaning of the word "case" was and I didn't feel comfortable to ask.

That night, I slept in that claustrophobic room. Early in the morning, I was awakened by a coarse and brutal sound from a deep sleep. In the public bathroom, I quickly washed myself and hurriedly sprinted to the train station. As the train arrived, I showed my ticket and went inside one of the compartments. The train was full of men. I could not find a place to sit. I stayed standing for two hours and slowly moved my way inside the cabin until, at last, I found a bunk bed. In fear of not losing my place, I didn't move away from my bunk bed until we reached Quetta.

In one of the towns where the train stopped, I saw that on the platform side of the rails, people sold tea, eggs, bread, snacks, samosas, and water. I hadn't eaten anything since lunch and was feeling voraciously hungry, so I brought my hand out of the window and one of the sellers gave me some eggs, bread, and tea. He handed me my purchases and before he could take my money, the train started moving. I was so glad to eat something, but I felt sorry that the man didn't get his money.

We were about one or two hours away from that station when a Pakistani man woke me up from my sleep. I woke up and looked down from my bunk bed. I saw the same vendor whom I bought food from. He was asking for his money. I was really surprised but pleased to see him appearing from somewhere. When I told the story in Quetta to others, they said that it is common practice for the sellers to give the material to their customers first and then get on to the train without paying the fee to collect their money from the passengers. Afterward, they take the return train back to their place. It was a practical and yet strange way to do business.

The Pakistanis often gave me curious looks and attempted to talk to me, but I didn't know their language so I just replied with smiles. Where I was sitting, the passengers of the lower seats wanted to play cards. Because they were short one player, they invited and asked me if I wanted to play cards with them. Because I didn't know what they were saying, unintentionally, I gave them a positive reply, nodding my head. They waited until I came down to play with them, but I didn't move from my bunk. After a moment, they called me again. I nodded my head as before, but I didn't go down. The third and fourth times, they said something I couldn't comprehend and then they all laughed. I looked silently towards them, not knowing why they were laughing.

The train stopped in every village and city on the way to Quetta. We rode all day long until, finally, we arrived in Quetta before sunset. Quetta was dirtier and much cooler than Peshawar. I gave the address to the rickshaw driver. I got lucky as the driver was from Kandahar's Farsi-speaking folks. Farsi-speaking people from Kandahar speak Farsi with a heavy Pashto accent. At the first moment, people think that these are Pashtuns, but later, it turns out that they don't know Pashto. The rickshaw driver told me that the situation in Quetta is very bad for Afghans, that the Pakistani police beat, imprisons, and harass Afghans everywhere in the city. Then he said, «You Agha Jan (an endearing tone) go directly to the committee. Don't hang around here and there. If the police catch you, they will make your life miserable." As he spoke, he pointed to the other side of the road where the Pakistani police stopped an Afghan man.

We arrived at the Jabha-e-Milli-Nijat Committee. I knocked on the door and the watchman opened it. I handed Habib's letter to him. He looked at the letter and I was sure that he was illiterate. He went in with the letter and opened the door after a few minutes, inviting me in. There were several rooms in the house. Several men were seen there. I walked inside the corridor and then, inside one of the rooms, a man who was perhaps sixty years old shook hands with me. I sat down in front of him. He began speaking, «You are sent by Hazrat Habib sahib. You want to go to Iran? I am the uncle of Amer Sahib (The Committee Head that the letter was addressed to). Amer Sahib went to Peshawar to see Hazrat Sahib. You have to wait until Amer Sahib returns.» This man was deputizing the responsibility of managing the office and house on Amer Sahib's behalf. The uncle was a talkative and kind man. I think that the family of Amer Sahib was also living in one of the many rooms in that Committee House.

A few days passed. The workers of the house treated me like a dear guest. The arrival of the Amer Sahib was still unknown. One day, the uncle drew me close and, intimidated by his fatherly voice, he said to me, "Every year from spring to summer, I go to Iran, and I am familiar with an Iranian family who cares for me. Being winter now, the weather is very cold at the moment and you won't survive the cold weather from Zahedan to Tehran. If you survive the weather, the smugglers and the thieves won't let you live." Then, he continued very kindly, "Take my words and go to Peshawar to Hazrat Sahib office. Work in his office till spring. Then, come back and I will take you to Iran myself."

His kind demeanor was genuine and I accepted his idea without any doubts. My mother had written my sister's address with a ballpoint pen inside my shirt pocket. She lived in Germany. My mother also had written the address of a very close friend of mine, Homayon (Also known as Homayon Uzbek). His nephew with the name Aziz. In Kabul, Homayon and I were hanging out together most days and nights. I knew his nephew who, at the time, lived in the city of Lahore. I waited a few days for Amer Sahib's return and when he did not return, I purchased a train ticket to Peshawar. This time, I knew that the tea and eggs were not for free. On the train, I was thinking about the behavior of Sibghatullah's women folk and found it strange that they didn't meet me while I was staying in their house in Islamabad. I remembered that my sisters sent a personal cigarette maker from Germany for my brother in prison. Because the guards didn't allow us to give it to my brother, I was using it. The days that Asma Mojaddedi would visit our home with her sister and mother, she would ask me to make cigarettes for them and to join them for a smoke together. Now, suddenly, I was a stranger to them. They didn't even remember that we used to live like close relatives.

I got off the train in Lahore and without knowing what I was doing, without the confidence that Aziz (the nephew of Homayon) was still at that same address written inside my jacket, I rented a rickshaw and using gestures and sign language, asked the driver to drop me off at the address.

CHAPTER FOURTTEEN

The rickshaw driver drove to a luxurious, upper class, pristine clean and refined area. He stopped in front of an incredibly beautiful mansion. I made sure that the address was accurate and then walked off. I heard my name from the roof of the house. I looked to the roof and Aziz was waving his hand, calling my name, "Khalil! Khalil!". He climbed down quickly, paid the rickshaw fare, and hugged me with all his kindness. Aziz was about two years younger than me with a short height, round face, and a beautiful smiley face. Back in Kabul, Aziz and I met once in a while, in reality, we didn't know much about each other very well, but his uncle was my very close friend. The young man who came down with Aziz introduced himself as a cousin. He took my bag and invited me in with delight. The Hall/Living room was immensely beautiful and luxurious, it had several other rooms were on the same floor and the owner of the house was living there.

We took the stairs to the second floor. There, we met a man who at first glance could be recognized as an Uzbek from our homeland. Aziz introduced him as his uncle, his father's brother. He shook my hand hesitantly, and then I was introduced to two children and a woman – his kids and wife. We got inside a room that was decorated with a sofa. Aziz started immediately recalling the memories of area Shahr e Naw and asked about Homayon and our other friends. After fifteen minutes, the uncle, seeing my condition, told Aziz, "Khalil needs a shower and change of clothes."

They showed me the way to the bathroom, provided me with clean clothes, and sent my dirty clothes to the laundry. They had prepared a very delicious meal for us. While eating, we told tales of Kabul and Shar e Naw and shared a great amount of laughs together. After about one month or more, I ate delicious food. Everyone was waiting for their American case (application to immigrate to America). After dinner, we went out to downtown Lahore, a vibrant city with many open shops. After exploring the downtown, we came back home late at night. They had prepared a clean and comfortable bed and I slept that night very comfortably, which was nice for a change.

The next day, after having a great, delicious breakfast, I said to them that it has been more than a month here and my family was unaware of my whereabouts. "I have my sister's address in Germany and I have to contact them to help me go to Germany," I said. Aziz took me to the post office and we sent a telegram to my sister's address. Later on, Aziz's uncle told me that the house belongs to one of the business partners of his brother Aziz's dear father. "The arrangement was facilitated by the many years of friendship with the Pakistani owner of the house and it is with regret to tell you that someone else isn't allowed to stay and sleep in this house. We are very sorry that you cannot sleep here any longer, but this is not a problem. We'll get you a hotel. You can come and stay with us all day and only at night go to your hotel for sleep until your travel arrangements to Germany are sorted out." Aziz was silent and seemed very uncomfortable.

Of course, I understood their position perfectly. I thanked them for all of their kindness and sympathy. With their help, I rented a room near the vicinity of the house and I went back to their house during the days with stories, laughter, and fun to share. After dinner, I went to the hotel to sleep.

I was sound asleep one day at five o' clock in the morning when Aziz and his cousin woke me up excitedly. Upon opening the door, Aziz happily told me, "Your sister rang last night and she gave me the address of an office in Peshawar where that some relatives of yours works there. Before the office closes, go to Peshawar and meet her."

He brought the rest of my clothes with himself. We went to the train station, they paid for my ticket, we said goodbye, and I was on my way to my next destination. I will never ever forget the love and kindness of Aziz and his family. Nearly a year later, I saw another friend in Peshawar saying that Aziz left for the USA with his family. I travelled to Peshawar through Rawalpindi. In Peshawar, I went to the address my sister sent me from Germany, feeling hopeful.

CHAPTER FIFTEEN

I got there at four o'clock pm. The office was near Sader (Center) Peshawar. The office was one-story with a huge lawn. Here was the Austrian Relief Committee where the Austrian government supported two Afghan refugee camps. There were a large number of employees, doctors, nurses, drivers, caretakers, and dozens of other duty-holders. There were huge differences in how this office was managed in every way between this committee and the 'Jabha 'Etehade Segana or Triple Union Committee. Perhaps their trustworthiness was the same, but while the Jabha 'Etehade Segana operated in a chaotic manner, the Austrian Relief Committee appeared civilized and organized.

There were several cars (vans and pick-ups) in front of the building. I knocked on the door and someone opened it. I told them the name of my sister's husband (Baqi Samandar) and the doorman let me inside. The corridor was fairly dark and cool, which was nice for the summer. Containers of supplies, food, and medicine were spread everywhere. A few foreign women were busy gathering things and two Afghan men were helping them. An Afghan lady was there too and she greeted me. After a short time, the door to one of those rooms was opened and a relatively tall man with a very handsome and well-dressed man with a stylish shirt, long hair Bruce Lee styled haircut, and a thin moustache who came from the Hazaras of Afghanistan entered the room and introduced himself. «I

Muammar Gaddafi

am Mohsen, one of the best and oldest friends of Baqi and your sister Soraya." He shook my hand sincerely.

Mohsen was not finished with his words before another person approached us and asked Mohsen if I was the brother of Soraya. Mohsen shook his head in agreement. The other man also greeted me with a friendly and sincere handshake and introduced himself as Naseem. Naseem looked like Muammar Gaddafi. Every time I look at him, I thought I was talking to Gaddafi. He said, "Baqi told us that you were coming. It's good to see you!" He asked about my journey from Kabul. After a few minutes, he said to me, "Until arrangements for your travel to Germany are completed, you will stay with Doctor Sediqa Mahmoudi. She left for home earlier today, but it's not a problem. We know her address and we will send you there." Naseem and Mohsen were the managers of the Austrian Relief Committee. They called someone over and asked him to drop me off at Mahmoudi's residence. As we were parting ways, they said if I needed anything, I should contact them anytime I wished.

We took a rickshaw to the center of Peshawar to a place called Qissa Khwani and Namak Mandi. I wondered why Dr. Sediqa Mahmoudi had rented a house in these slums of Peshawar. The rickshaw stopped on the main road and we got off, we walked through narrow and crowded alleys until we reached a flight of stairs to climb. Once we reached the top, we walked to a three-story building. As we went up the stairs, on every floor there were Afghans living in rooms. In this house, Doctor Sediqa, her mother and two grandchildren of the late great man, Dr. Abdul Rahman Mahmoudi (Founder of the Afghan Communist Party), lived here. The grandchildren were Doctor Jawad Atayee and his sister. Dr. Jawad Atayee's father was well-educated, kind, and had a great personality. May his soul rest in peace. He was my father's great aunt's son and the son-in-law of Dr. Abdul Rahman Mahmoudi.

We knocked at the door. The older sister of Dr. Jawad Azada opened the door. She immediately recognized me and gave me a handshake. She knew the man who escorted me here too and greeted him kindly. She warmly welcomed us inside. This house had a corridor, two rooms, a toilet, and a kitchen. We entered a room to the right of the dining hall where Doctor Sediqa, a woman in her late thirties with a skinny and slender body, greeted me with a handshake. Her mother was sitting beside her and I bowed down to kiss her hand. I'd never seen Sediqa and her mother before and I didn't know anything about them. To my wonder, Doctor Sediqa knew me and told her mother that I am Yassin's son. My companion left after a cup of tea.

The questions began, about where I was and what I did and so on and so forth. During the conversation, they told me that Jawad was out of the house and returning soon. I had seen Dr. Jawad several times in our respective homes. He was two years older than me. Although we were from the same extended family, I didn't know Dr. Jawad that well. Dr. Jawad was the only son of his family. He lived amongst five sisters. Being the only son, Jawad was loved by every member of his family and each member of the family always made a fuss about his health, wellbeing, and security, especially in the city of Peshawar which had animosity towards Afghans.

Jawad and I met and we forged a closer friendship here. One day, we went to a park called Baqi Shahi where a social gathering of Afghan refugees and Mujahideen congregated. We weren't supposed to go there, but we often did and played soccer with the others. One time, Dr. Jawad's family sent him new pair of pants from Afghanistan. When we played soccer and Jawad fell, his new pants were torn. In order to protect ourselves from his family finding out about this and to avoid telling them where we were, we made up a story that when we were leaving the bus, a biker didn't notice Jawad and hit him with his bicycle. We said when Jawad fell down, his pants were torn at his knees.

Dr. Jawad was a young, sober, cheerful, and kind man. Best of all, he was extremely well-mannered. We always laughed from morning to evening. He used to smoke a cigarette in secret often. In one room, ladies slept, and in the other room, Jawad and I had beds. As soon as the lights went off, we would light a cigarette or we'd go outside to smoke by making excuses to his family, such as buying groceries. Jawad sometimes spoke to me about literature and poetry.

Outside in the market, anything was traded. Sheikh Mohammadi, a tribe from Eastern Afghanistan known for being crooks and swindlers in trading, controlled the entire market. From sunrise to sunset, they were fooling people to depart with their money, especially the Pakistanis. It's a well-known rumor that the Pakistani men suffer from sexual inadequacies and Sheikh Mohammadies sold them any rubbish, tricking them into believing

Peshawar city

that it solves their sexual inadequacies. Once, I fell foul of their trickery but not in the way the Pakistani men fell for it. I was tricked to purchase a wrist watch that was more expensive than a shop.

Outside of Dr. Jawad's house, there was too much noise, but inside of the house, it was calm and quiet to the point that no one could hear even the sound of mosquito flying. At night, Dr. Sediqa and Dr. Jawad's sisters were learning English with Dr. Jawad's tuition. Some nights, they would start playing card games for the sake of my joy. We played "king and thieves". Dr. Jawad and I gave more cards to his grandmother and she would always feel lost and become thief. We'd laugh every time.

I shaved my beard after a while. Dr. Jawad was preparing to study medicine. Sometimes he'd need to go to Multan to sort out his entry to a Medical School. He was also looking to get an entry for a Medical Degree in Peshawar. During this time, I'd be left alone with his grandmother in the house. I'd have nothing better to do than daydream. I looked out the window and saw airplanes in the sky. I imagined speaking enthusiastically to the passengers of the plane. I dreamed of leaving Qissa Khwani at the earliest opportunity to join the civilized world.

At the end of the alley, there was a mosque that was opened like a factory 24 hours a day and produced little Imams/Mullahs. In the mornings, the mosque became a madrasa (religious school) with children sitting on the floor wiggling like worms, reading and memorizing Quran in Arabic like parrots without knowing the meaning of it. The misfortune was that all of Pakistan's mosques had loudspeakers and twenty children would read the Quran for hours through loudspeakers, which was very annoying and irritating.

I lived with Dr. Sediqa's family for a few months. During that time, they treated me like a family member, always with love and affection. In that house, I experienced the true meaning of love, kindness, intimacy, honesty, brightness, and compassion. However, I was mindful that I felt I could not burden my presence any longer to them. I was also hungry for adventure.

As luck had it, one day Dr. Sediqa told me that Naseem wanted to see me. I went to the Austrian Relief Committee. Naseem told me, "Your sister and brothers have sent you two-thousand German Marks for your travel

Me Dr. Jawad and second cousin Homayon

expenses to Germany. I will pay you the money in one or two days." The next day in Peshawar's Sadar Bazar, someone put their hands on my eyes from behind (an Afghan tradition where you need to guess the person doing it). I said a few names but all were wrong. Finally, he pulled his hands back from my eyes and I turned around only to see my best friend, Hayat. We were like brothers. We hugged each other for a long embrace. We couldn't contain our happiness. Hayat was living in the Green Hotel with his brother. He took me there to meet his brother, Rahmat.

CHAPTER SIXTEEN

Inside the narrow lobby of the first floor, we climbed the stairs and walked into one of the rooms where I met Hayat's brother, whom I never met before in Kabul, and another young man. His name was Gran and he was about nineteen years old, strong, tall, curly haired, and he had only about twelve teeth in his mouth which were rotten. Gran was also trying to flee to Germany. Hayat and his brother had applied for US immigration visas. They asked me about my plans. I told them that my family sent me two-thousand German marks that I intended to use to fly to Germany.

The Green Hotel was located in the Sader area of Peshawar along the main highway street with a large balcony facing the road. This hotel was the center of the Afghan breaking news station[6]. Often, stories of most interest were: who was imprisoned, who left successfully to other countries, who is being caught in the airport, which human trafficker or smuggler was successful in his job, which trafficker has ran away with the people's money, who became Commander and prepared a group to take arms to Afghanistan, who's immigration visa case is rejected, what is the price of travel to Germany, France, America, Canada, which fake passports were good, which country accepts Afghan refugees, and hundreds of other news were speculated during the day.

The Green Hotel was the choice news station for Afghans. When they arrived in Peshawar, all Afghans from all walks of life were almost mandatorily spending a night at the Green Hotel. If the situation was

---

[6] Better to say, the Green Hotel was the Facebook of its time, where you got all sorts of fresh news about love, war, prosperity, frustration, etc.

Me with Rahmat in Green hotel

Me and a friend Weis Hadi

filmed every day in the Green Hotel, the following scenes one would have seen would be dramatically akin to those in a movie. There would be an Afghan family with three children half-asleep and half-awake would be guided by an adult, papers in his hands, tired from traveling to Islamabad from visiting the American, Canadian, Australian, French, or German embassies with no good news of their ability to leave. Instead, simply told to wait. Another Afghan, who had been looted by a smuggler, would now be looking for that smuggler with drowsy eyes, an open mouth, confused look on his face, and still he doesn't know when he would be allowed to move to a foreign country.

Another family would be there too, expecting their families who were living abroad to send the money to them. A week would go by for that family and then they were visiting the shop where the money was supposed to be transferred to, only to find the money hasn't arrived yet. They were going from room to room in hopes to find or borrow from someone to pay their bill at the hotel. On the contrary, two young Afghan boys, cousins to one another, would be there to receive their money from abroad. They tailor-made clothes with a Pakistani tailor, ate food in nice restaurants, were smoking expensive cigarettes, ogling at Peshawari women from their hotel balcony while being wrapped in purdah of Islamic outfits where only their eyes could be seen.

Another Afghan had the ingredients in their hands to prepare dinner for the evening. From most of the rooms, there was a whistling sound that emanated from the Iranian pressure cookers. A few Afghans gathered together in a room and spoke about politics. In another room, two Afghans are drowned in the strategic thinking of playing chess. In another room, some visitors have arrived to meet with dear ones. In each room, there are three to six Afghans living together.

In the Green Hotel, one could find any type and shape of Afghan from all walks of life. Whether it was a carpenter, teacher, student, Mujahid, merchant, mechanic, chef, government agent, or engineer, all kinds of Afghans were there. However, there were three Afghans there were the most interesting, famous, and unique amongst all.

Farid was known as Farid Reesh (bearded Farid) whose beard was nothing special, but he was famously known as Farid Reesh and he was like all the other bearded people of Peshawar. It was unclear for what reasons they called him this nickname. Farid was talking very quietly working on finding passport to smuggle people out of Pakistan. He was very kind and hospitable. His room even lacked a lock. Everyone from everywhere would spend the night in his room, even if he was not in the city. His room served his fellow countrymen.

The next notable person was Nassir, a Farsi speaker from Kandahar who had two titles: Nassir Kandahari or Nassir Hazara. His face was similar to the famous Hollywood actor Charles Branson. Nassir was like the Godfather. Everyone and everything was in his control. He had friends from every class of people. He was illiterate, but he also had educated friends. He controlled the rooms and hotel laws. He was very generous hospitable, and always had guests staying in his room. He cooked with great joy. He wouldn't put up with nonsense. He didn't like forceful and pretentious people. His word was accepted and honored by his roommates and his friends.

The last famous person was Baba Azim known by Azim Tula (Azim flute). He was from Chardehi Kabul. Azim was a very thin, tall man with long hair and a beard. He introduced himself as a Mujahid, but it was debatable whether he was a Mujahid or not. Some people said that since the day he left Kabul; he had never gone back. Others said that in the summertime's, he goes to Paghmān and returns back to Peshawar in the winter. I don't know where he was living, but almost every night he was in the balcony of the Green Hotel until late. It was fun for everyone to be around him. All the Afghans of the Green Hotel and even foreigners loved him. Baba Azim had three unique characteristics: he was constantly smoking hashish, playing the flute, and using catchphrases and particular words in his own way. Over time, every Afghan who got to know him gradually started to use his catchphrases.

Baba Azim put the hashish over the hookah and would dangerously suck the water pipe so that the fire would come out on the release. When he puffed out the smoke, he would say his particular hashish catchphrases. He even taught these words and phrases to foreigners. Waaah waaah We smoked hashish and got to know the highest heaven of truth. We did not expect this from this weak grass **
Hayat told me that I met him at the right time because tomorrow a large number of the Green Hotel inhabitants are moving to Rawalpindi. He told me to go with them until my travel to Germany was finalized. I accepted his

Nassir Kandahari /Hazara

Baba Azim

request. I slept for the last night at Dr Jawad's house. The next morning, I said my farewells, expressed my gratitude for them, and departed for the next phase of my journey. A large group of Afghans, including two families and numerous young unmarried men, moved to Rawalpindi.

** About twenty-five years later, in 2004, when I was in Kabul and organized a music night, Baba Azim came with one of my friends, the same face and pastures but with grey hair, the same flute but he didn't repeat the slogans anymore. Apparently, he left all that in the era of jihad in Peshawar. I introduced myself he didn't recognize me. We did mention Peshawar and the Green Hotel. I was immensely happy watching him being alive. This was the famous Green Hotel, which gave a glimpse into the combinations of Afghan sorrows, uncertainty, worries and despite all that their fun and laughter as an antidote to miseries of being displaced .

CHAPTER SEVENTEEN

We travelled to Rawalpindi with a small convoy of mini vans. We had fun the entire drive. Everyone knew each other, except I was the only stranger to everyone. Finally, we arrived at Rawalpindi's Shabnam Hotel in the area of the Liaqat Garden. Shabnam Hotel had one gate as an entrance. Alongside it, there were two shops, one of which was selling milk, cookies, and some other small grocery items. The other was a barber shop with four bathrooms at the back of the shop. Shabnam Hotel was constructed of two stories in a square shape. In the middle, there was half-an-acre of a yard and all four sides of hotel rooms had windows facing into the yard. Similarly, the second floor had a corridor to the left, with three rooms, the rest and the more rooms were around it. The lower floor of the hotel was used by Pakistanis.

One of the rooms on the second floor was booked by a group of five friends: Khaliq Khaliqi, Mahlem Waseel (Teacher Waseel), Ghafor (nicknamed Ghafor Souz or Green-eyed Ghafoor), Aziz Chersi (Aziz Hashish), and Zabi Gottak (Zabi Short), who was the cousin of the Yassian Sheer Peera, head of the Afghan National Soccer Team. In the same lobby, the opposite room was booked by «engineer» Obaid. (God knows if he was an engineer or not). He was living there with his wife and two boys. His second child was born in the same hotel and they named him Khalil. The third room was also booked by Afghans. Out of the corridor,

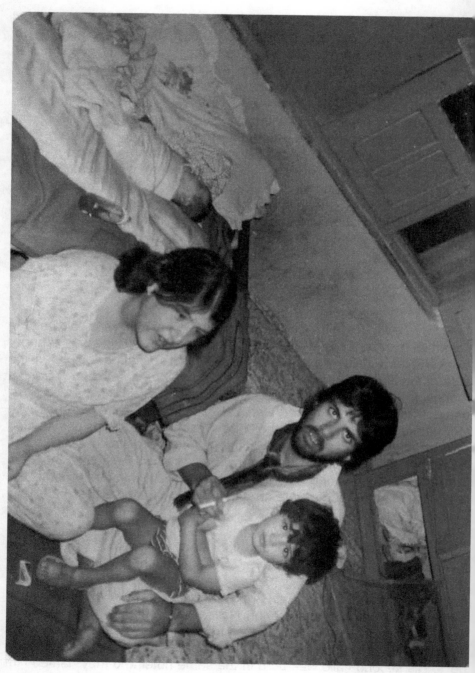

Me with engineer Obaid, wife and children (new born, his name is Khalil)

Me with Hayat and Gran

there was another room with two brothers, one named Wahid who had a wife and little daughter who was adorable like a doll, and his elder brother Abdul Rahman, who was known by the nickname Mamoor (Clerk). In the next room lived Hamid and Mehdi from Iran. In the fourth corner room there was Nasir Hazara/Kandahari. The next room was taken by two cousins, Salem and Anwar, who were intending to go to Europe. Another room was booked by Hayat, his brother Rahmat, Gran, and me. Another room next to ours was booked by two younger brothers whom I knew from Kabul. Most of the rooms in Pakistan's hotels didn't have a window.

A number of Afghans who were with us were all trying to go to the Western countries. There were some who were waiting to return to their home, hoping that the situation in Afghanistan would improve soon. One of the reasons that these residents of the Green Hotel came to Rawalpindi was because of a man named Wakil who was a good friend of Nassir Hazara. Wakil lived on the opposite side of the road in a hotel more decent than the Shabnam Hotel.

In those days, I spent my last few hundred rupees from all the money I brought from Kabul. After a couple of days, Hayat took me to Para Chinar, the border of Pakistan and Afghanistan, to his sister's house. The interesting thing I noticed was that a number of Afghans had land and property in Pakistan. They went to Pakistan in the winter and returned to Afghanistan in the summer. Other Afghans called them «the sheep of two maids». Hayat's sister and his brother-in-law had land and property in Pakistan. We stayed there for two nights and then left back for Rawalpindi. Two weeks later, Hayat went to his sister's house and never returned for some reason that I never knew.

As the days went on, I got to know my new friends. Nearly everyone knew that I would be leaving for Germany and that my family had sent me two-thousand marks. Only trouble was that based on the current rates of smugglers, I was only a short three to four-thousand Pakistani rupees. Once I had that amount, I could go to Germany.

At that time, everything in Pakistani markets were Pakistan's own products. Only in frontier provinces by the border of Afghanistan and Pakistan were there foreign products being smuggled from Afghanistan into Pakistan that were available for purchase. A large number of young Afghans, including Hamid Irani, Mamoor, and several others from the

Me with Mahlem Waseel

Khaliq Khaliqi, Mahlem Waseel, Ghafor, Aziz Chersi, Zabi Gottak, Hayat, Gran and the rest of crew

Abdul Rahman,(Mamoor)

Shabnam Hotel, started the illegal business of purchasing goods in the frontier province, like children's toys, clothes, soap, watches, steam irons and hundreds of other objects, and took them to Lahore where they sold them. To make up the shortfall in money, it was advised that I should also go with them on one of their trips and purchase some products to resell in Lahore. They told me that by doing this, I would surely be ready to go to Germany.

Rahmat and I went to Peshawar and we spent most of the night with Baba Azim listening to him play his flute as usual. When we got tired, we fell asleep in Farid Reesh's room for the night. The next day, I went to the Austrian Relief Committee to get my money. I received about twelve thousand rupees. We went with Rahmat and two others to Parachinar to buy products. I bought soap, baby pacifiers, baby milk bottles and some other things.

We travelled with our merchandise back to Rawalpindi. On the way to Peshawar, buses were checked by police at Attack town by the Attack bridge, which is the boundary where Northwest Frontier Province ends and the Punjab Province begins. You had to keep the goods concealed so that they didn't get noticed by the police or you'd have to bribe the checkpoint police, or worse, as it happened very often, the checkpoint police would capture the goods. As luck had it, we brought our goods back to the Shabnam Hotel without detection. The next day, Rahmat, Mamoor, and I went to Lahore to sell our goods. Once there, we checked into a hotel and sold our merchandise. The materials were sold and we made a decent profit. Now, in my pocket I had about fifteen-thousand rupees, enough to travel to Germany.

CHAPTER EIGHTEEN

I heard that the Islamic Republic of Pakistan had very large brothel houses and the most beautiful one was in the city of Lahore. Based on my first experience of the place in Lahore, it was not far from the truth. The area where this brothel was located is called Hira Mandi (Jewels market). After selling our goods, Mamoor and Rahmat told me let's a have walk towards the Hira Mandi but to keep this as our secret. I agreed and we headed there.

Me and Rahmat

According to Islamic law and the law of the Pakistani state, the existence of such places was prohibited. In Hira Mandi, only occasionally, a group of policemen patrolled those roads. Every now and then, an unlucky passer-by got their wrought. They would charge the unlucky punter with some stick beatings on his legs or on the back and then they thought that they have done their duty according to the State and Shariah laws.

We went to Hira Mandi. The chambers were slightly higher than the surface of the Earth. In the chambers, dancers and musicians were waiting for the punters to come by. Close to the chambers were money exchangers sitting ready to exchange big bank notes into smaller ones of one, two, five, and ten rupees. We chose a room where a dancer was performing. We exchanged an amount of money into small bank notes. The door and the wooden windows were closed to protect our privacy and avoid uninvited attention. The music began. The young pretty girl was singing very beautifully. She sang some songs from Indian films and danced at the same time. We enjoyed sitting on the mattresses while hearing this beautiful music sung by this beautiful dancer. She sang two or three more songs. We enjoyed the exotic experience and finally got out of there.

A pimp approached us and said that he had a better dancer. Rahmat told him that we will go there with one condition: if he could find liquor for us. The pimp said, "No problem." We got inside a house. The pimp directly guided us to the second floor. There, a woman who appeared to be a 'Madam' in charge of the house joined us. The pimp told her our request for liquor. The woman went out and returned a few minutes later. She told us that she had sent a person to bring us liquor. She warned us not to be too adventurous, to drink sensibly and as quickly as possible so we don't attract the attention of the police. We agreed. They served us metal mugs and after a few moments, they brought us the bottles of an orange liquor. They told us that this liquor was made in India and that its much better than Pakistan's liquors. They filled the metallic mugs and gave us our drinks. The pimp and the woman looked out occasionally of the window to the road to watch for police and they rushed us to finish the liquor quickly. The liquor tasted more like a spiced-up water than liquor. The bottle of liquor was finished quickly and we were drunk like a thousand Russians. The Madam locked the brothel house doors, made the environment private, the tape recorder was laid in the backyard, and pretty women were dancing, flirting, and playing with us. This was all being paid for by the profit of the goods we sold, blowing out from my left and right pockets. At one point, Rahmat told me to give him my money so that I couldn't spend it all. I gave Rahmat about ten-thousand rupees. We stayed there until early morning. Eventually, we became tired and sleepy. We decided to go to the hotel and get some sleep. After leaving the brothel house, we lost Rahmat. He vanished in that crowded place. I don't know how Mamoor and I got to the hotel, but as soon as we were there, I threw myself on the bed, fell asleep, and didn't know anything.

Sometime early in the morning, I felt that someone was shaking me vigorously. I could hardly open my eyes. I saw Rahmat with a distressed face standing over me and saying that I should go outside with him urgently. He told me that a policeman stopped and searched him. The policeman asked him where he got all that money from. The policeman took the money and sent him to bring me. I put on my shoes in a rush and went outside, but there was no policeman to be found. Rahmat was running from street to street, but all in vain. The police were gone and we returned back to our hotel with empty pockets, utterly exhausted.

Rahmat and I didn't own another rupee. Mamoor paid for our trip to Rawalpindi. On the way back, we promised each other to keep this episode as our secret and if someone asks about the money, we will tell them that the police had confiscated the smuggled goods, including the money. Ever since, this secret has been like a weight on my shoulder[7].

When we arrived in Rawalpindi, Rahmat promised that he would pay back my money. He went to Sayed Ahmed Gailani[8] to borrow money for me because he had family relations with him. I told him not to count on family relations with rich and powerful people. I told him about the family kindness of Mojaddedi, but he was not listening to my story. On the very first day, a large number of the Shabnam Hotel residents knew that I have been looted by the police. My poverty and misery began again. I had no money at all. I had no money for rent. I had no money to go to Europe. My sisters and brothers raised and borrowed two-thousand Deutschmarks with difficulty for my salvation, but I ruined their two thousand marks in one night. After that, every six months, they would send me one-hundred Deutschmarks, which was hardly enough to buy me cigarettes.

---

[7] This is the first time that my family and friends will know the truth about my disgraced experience in Lahore.

[8] Founder of the National Islamic Front of Afghanistan

After a week, Rahmat returned and gave me one-thousand rupees, saying that Gailani's family didn't give him more than that. I figured that the Gailanis and Mojaddedi's were similar. "Don't worry," he told me. "I'll do my best to find the rest of your money[9]." He stayed for a few hours with me and then left to return to Peshawar.

---

[9] I didn't see him for a very long time and neither did I receive a penny from him.

CHAPTER NINETEEN

Gran and I stayed alone in the room. Nassir Kandahari's room was always on the move. He had a small tape recorder that he'd often play Mohammed Rafi's songs (Bollywood singer). In his room, different characters were coming and going. They were people who travelled around for work, such as Abdullah Ali Ahmadi, a student of Engineering in Kabul. In the other spectrum there were all types of misfits, smugglers, and so called Mujahideen hence why Nassir's Iranian pressure cooker was always cooking.

Residents of other rooms were also just eating, gossiping, speculating on daily affairs, and all attempting to get out of Pakistan. There were some people who ate outside mostly, including my roommate Gran. There was a funny and fascinating character named Wakil. He used to sneak up quietly once or twice a month from the stairs and, just for fun, steal Khaliq's pressure cooker with food before running back to his hotel. The guys would get angry and go after Wakil, beat him, and bring back the pressure cocker. Among all of them, I became closer to five guys: Khaliq, Teacher Waseel, Ghafur Souz, Aziz Charsi, and Zabi Gottack. These guys and several other Afghans were receiving a monthly allowance through UNICEF. The United Nations paid very little financial support to Iranian refugees in Pakistan, and because they had lived in Iran, were familiar with the Iranian accent, traditions, and customs, and because they had enough information about the streets and the rest of Iran to pass a test to prove being a local Iranian to receive the meagre UN's monthly stipend. I became a closer friend with Khaliq.

Khaliq was a relatively short person who was hyper, sharp, thoughtful, alert, sharp-witted, intelligent, well-read, interested in chess and Carrom board. He was a gracious, loyal good friend[10]. Khaliq soon realized my frightful state. He told me, "Khalil, you can eat with us until your financial situation gets better." I still don't know how he convinced his four roommates with regards to this. Khaliq sometimes bought cigarettes for me and would never hide his cigarettes from me. Most importantly, Khaliq encouraged me to read books. He brought a box of books from Iran that he guarded jealously from others. For the first time in my life, Khaliq managed to awaken an interest in reading and the enthusiasm for literature. I can confidently say if it weren't for Khaliq's encouragement, my spelling, grammar, and the slight literacy skills I had would have diminished in my enjoyment of cool late nights spent on the roof of the hotel. At dawn, everyone was asleep while we were

---

[10] Our friendship has remained strong to this day

searching the ashes for the fag ends of cigarettes, searching every room in hopes to find fag ends. We would then roll the tobacco from the fags into paper and made our own cigarettes from wasted ends.

One particular day. I didn't have any money to buy cigarettes and I badly wanted to smoke. I remembered an experience from Kabul, once contrary to our customs, where I smoked in the presence of my father. My mother was horrified and said, «Smoking is a bad act that you should not take up, but now that you smoke, until you are under my shadow, I will give you money to buy your cigarettes so that you do not beg others for cigarettes." Remembering this made teardrops roll from my eyes because I needed a cigarette and my mother was not there to give me the money for them.

We slept almost all day. Unlike my terrible economic situation, Gran was well off economically. He received help from everywhere. He occasionally used to eat with us in Khaliq's room for free, but he mostly ate in restaurants with a bottle of Coca-Cola to himself. Someone told him that he's drinking Coca-Cola excessively. Gran told that person, "You should know and understand that I don't have good teeth and my stomach does both tasks of chewing and digesting. Coca-Cola helps that process." Gran eventually managed to go to Germany. Sometime later, he sent a letter to us addressing everyone. He wrote to me and Hayat about how Germany is a beautiful country. He mentioned that the price of an uncooked chicken is only four Deutsch Marks and the price of a cooked chicken is six Deutsch Mark[11].

I was totally reliant to living with other peoples' meal time and what they ate. I was often waiting with a growling, empty stomach until someone would prepare their meals and offer to share with me. For too many days I just looked at cookies and cakes through the glass windows of shops and imagined how they'd taste. With all that misery and misfortune, I also became ill. Afghans were not used to the extreme heath of Pakistan. Consequently, due to extreme heat, moisture, lack of washroom facilities, and a disregard for health, the penile area of most Afghan men developed a painful, scalding condition which itched and burned constantly for a relatively long time. You could point an Afghan man by the way he walked, like a cowboy in Western films.

---

[11] Years later, I saw Gran in Germany. His face had changed a lot. I didn't recognize him but he recognized me. This time, his mouth was full of teeth.

CHAPTER TWENTY

One after another, my friends flew to Europe, the United States, Canada, Australia, and other countries. The number of my friends gradually decreased while my frustration increased. There were constantly new arrivals of Afghans continually coming into Pakistan from Afghanistan. None of this affected Khaliq and Nassir Kandahari; they both continued to be kind, generous, and helpful to everyone alike. One of his friends, Taher Kandahari, had been addicted to heroin. He would tie himself up in the room to endure the addiction withdrawals. His cries were heard from far away. When he was out on his own, he would steal money from his friends to buy drugs.

Another famous personality was Haji Shlaghm (Haji Turnip). He hated turnips. Every time he visited Nassir's room, after receiving great hospitality, Nassir or someone else would leave a turnip inside the room and everyone would laugh as Haji Shlegam would leave the room while cursing everyone's mothers and sisters. Another character was Hamid Irani. He was a thin man with curly hair and a very strong moustache. He was shrewd and canny, an Iranian in all senses, who escaped with Afghans from Iran. He quickly learned how to smuggle humans and goods, leading him to become very rich. Every now and then, he sent out people with fake passports obtained from his fellow smuggler, Aziz Atash (Fiery Aziz). Hamid Irani would try to cheat and argue with Aziz. Aziz would get worked up and chase him everywhere while Hamid would run away. Finally, one day Aziz got hold of him and asked him for the money or to give him back all his fake passports. Hamid gave him his money.

After Gran left, my problems increased with renting the room. Someone brought a friend to the hotel who was looking to share a room. His name was Assad. His nickname was Assad Jani (Criminal Assad). Assad Jani earned a living from a Polaroid camera photographing people in Islamabad's parks. Assad was a complex character. He was professional and extremely skilled in the carrom board. We also had a strong soccer team with the Rasoul, an ex member of the Afghan national team, professional soccer players such as Zabi Gottak and Sakhi.

Among our friends, we had the best chess player by the name of

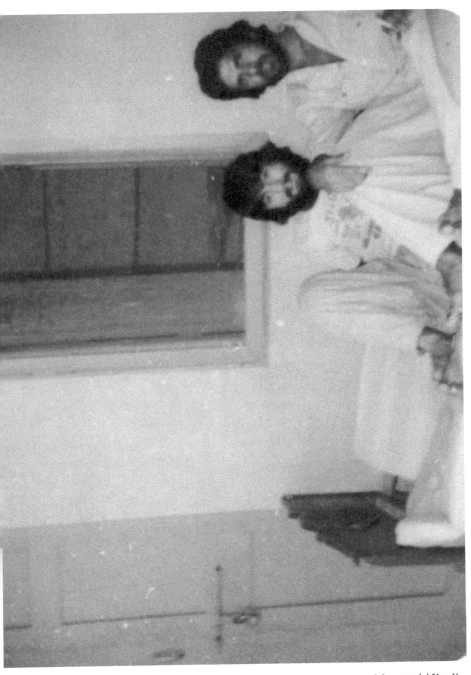

Me and Khaliq

Our soccer team (Shabnam Hotel)

Haji Din Mohammed (Nassir Kandahari's roommate). He was almost unbeatable in chess. He was known to even play without the Queen, Bishop, and Rook, and still win. We used to go to other hotels with our team where Afghans held chess competitions and we always won our matches against them.

We also passed time by having movie nights. Smugglers, such as Hamid Irani and others who traded well, invited us to the movie night. They rented a TV and video players for the night from Pakistani shops. The package came with three video tapes: a martial arts movie, a Bollywood movie, and a porno. More than twenty of us sat crammed beside one another in a small room. First, we watched the porno because of the fear Pakistani police would catch us and also because of high demand for such rare movies, the guy from the video shop would be waiting outside the room and as soon as the movie was finished, he would take the video and vanish quickly to take it somewhere else to be watched by others. Then, we would look at the other two films until sunset before falling asleep at night.

During the month of Ramadan, Pakistani cities undergo a complete transformation in the evenings. From the time the fast breaks until dawn, everywhere becomes flooded with lighting, the sounds of music, the hustle and bustle of restaurants, juice vendors and their customers. There are live concerts of qawali (a sufi or mystical singing). Most people were fasting. I didn't have food so fasting was a good excuse for me. I was equipped for Ramadan prayers as someone already taught me a basic prayer known as Al-Hamdu and three Qull[12]. Some days I used to go to Islamabad to visit friends and eat with them.

One day in Islamabad, I met an old friend from Kabul. He was looking great, smuggling drugs for a living. He asked me how I've been doing. I explained my predicament and mentioned that I got into bad financial problems. He suggested that I should work with him. I thanked him and said I would need time to think about it. I thought about it all day and night. Although my circumstances were very unpleasant, I didn't want to be involved in smuggling drugs and thus, politely declined his offer.

---

[12] The opening and the last three Surahs of the Quran. Because they are short but powerful, they are generally the first verses of the Quran that Muslims memorize to meet the minimum requirement for the five daily prayers.

Our soccer team (Shabnam Hotel)

Me, Mahlem Waseel, Ghafor Souz, Zabi Gottak, Asad jani and some Pakistani friends

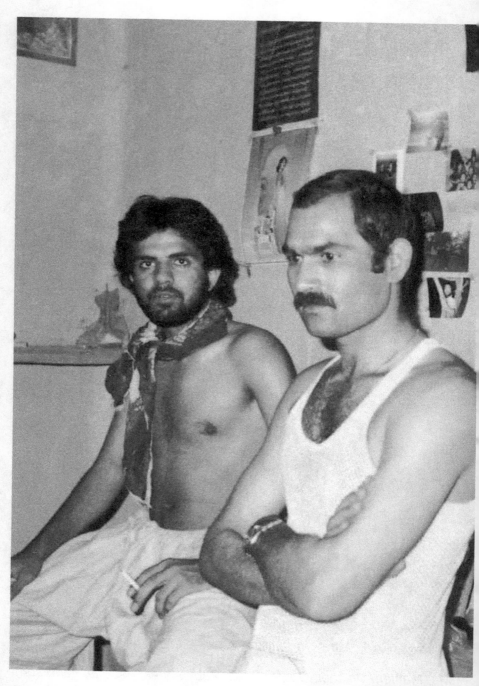

Me and Mahlem Waseel

Islamabad friends

Eventually, I decided to go to the United Nations office and introduce myself as an Iranian. As long as I live in Pakistan, I will survive with the help of the United Nations. To prove my being Iranian, I had to undergo a test. I was tested by a harsh and strict Afghan woman. The night before the test, I took lessons from my friends who spent time in Tehran. They coached me on perfecting the Persian accent and taught me about Tehran's landmarks, neighborhoods, and famous streets. The next day I was dressed in stylish clean clothes as I made my way to the United Nations Office in Islamabad. They called my name and after just a couple of questions, the examiner told me that I am an Afghan. I tried to argue but she said I better leave immediately before she informed the police. The door to my hope was closed along with the door to the office.

Wakil knew a man who sold ice. This man hooked him up with a liquor seller and sometimes Wakil shared a glass with me. Most guys were smoking hashish. I hadn't done it yet because I believed that hashish would paralyze me and cause me to lose some part of my body, so was not interested in hashish. Although, back in Kabul, my friends Wais Hamed, Homayoun and Khawja Hamid sometimes smoked. I was always too afraid and stayed clear from it.

When someone intended to travel to Peshawar for safety and company, he would take somebody to accompany him. I was often that companion. Thus, we'd go to Peshawar and besides whatever task they were doing, we'd also visit friends and slept in the Green Hotel in someone's room. On one of these trips, Baba Aziz Tula was playing flute on the Green Hotel's balcony while a number of people gathered around him. He told his tales of Jihad while smoking hashish. Someone got liquor, a group of Kabuli men quietly snacked to Farid Reesh room to join the liquor party including me. One of the guests filled the cigarette with hashish. Then, he gave it to Baba Azim to light it and he'd pass it down the circle. The round started and eventually reached me. Without taking a puff, I passed it on to the person after me. Suddenly, hell broke and every one was protesting that I cannot break the circle. Apparently, there is a code and a lot of superstition is attached this code that breaking the circle brings untold miseries to the group and is regarded as a highly anti-social and unethical behavior. Hence, I got my first smoke of hashish. I didn't get high on it, but I thought soon I would lose some part of my body. Fortunately, nothing of the sort happened. After that night, I also joined the crowd and occasionally smoked hashish with others.

CHAPTER TWENTY ONE

My days were passing by slowly, filled with sleeping, hunger, and misery. The first time I received information about my family was from one of my neighbors who used to come and go to Pakistan often. My mother sent me some clothes in a package. It made me so homesick and left me missing my parents with such intense pain that I cried like a baby for hours. This was the second time I was crying uncontrollably from my heart; the first time was Koh e Safi (Safi mountain) when I was left alone at dusk, not knowing where I was heading and what dangers were awaiting me. This time, I was crying while reading my mother's letter. My friends, especially Khaliq, showed their sympathy in kind words and tried to make me feel better.

The winter was approaching and the weather was slowly getting cold. Sending and receiving letters to and from Germany took months. In one of my letters, I made my brothers and sister aware of my despondent state. I received a reply letter from my sister who wrote that one of Baqi Samandar's (her husband) friends had a carpet washing company in Rawalpindi and that he needs help. He needed someone's to live with him and work with him in his carpet washing business. One day, a thin, limp man with a bent back who was about thirty years old came to visit me. He introduced himself as Ashraf, a friend of Baqi Samandar. He was a relatively tall man with brown hair. He spoke very slowly and hesitantly, forgetting the correct words in most of his sentences. From the state of his speech and mannerisms, it was obvious that he was ill. I didn't have any other choice, so I collected my luggage and said goodbye to the guys before leaving with Ashraf to his home.

Ashraf lived in a room on the second floor of a house in Rawalpindi. He lived with his father, mother, and younger brother. Ashraf's brother was tall, girly- looking, and a few years younger than me. Ashraf had two sisters; one was married and the other was single. Both sisters worked with the Austrian Relief Committee. He had another brother that he didn't talk much about him. His sisters, especially the single, sometimes came to visit. Although Ashraf lived on the second floor, most of the Pakistani homes were shaped like three apartments; the first floor had a yard, the second floor also had a shared patio, and the third floor had a patio on the roof of the building. Ashraf's apartment was pebble-dashed.

Baqi my brother in law

Ashraf Brothers

Ashraf's carpet washing company was interesting. During all the time I was around him, he only had one carpet from someone to wash. In the middle of the square-shaped yard, he had nailed the carpet with four knuckles and raised the middle of the carpet with wood so it could be stretched. He didn't touch the carpet at all because he was sensitive to dust, or at least he claimed so. Every day, he argued with his younger brother about washing the same carpet. His father was busy writing outside in the sunshine during the mornings after breakfast. My presence there was a bit precarious. His mother used to hide her face from me and occasionally mumbled somethings. His father didn't like to talk to me at all. Occasionally, I returned to the Shabnam Hotel to visit my friends, have a change of location, and smoke of hashish.

Winter began. One day, I received a message from Shabnam Hotel that my brother Farid has come from Afghanistan. I went to the hotel where my brother was waiting for me in Khaliq's room. We hugged each other, crying and laughing in our reunion. He told me all the news from Kabul, about our father, mother, sister, nephew, and brother Zalmay who was in prison. Now, only three adults and one child remained in that big family. I noticed that he was suffering from pain in his leg and a part of his ear was missing. I concerningly asked him about it. Farid began speaking in a sad-angry tone, "The puppet government of the Khalqi and Parchami fired me from my job and forced me to join the army. During one of the expeditions from Mazar to Kabul, a mine had exploded under a vehicle and left me injured. I was transferred to Kabul in a very bad condition, which resulted in a wounded leg and a piece of my ear missing. I spent some time in the hospital and eventually I was discharged from the hospital to go back to the military. Fortunately, I managed to escape my post in the border of Torkham and made it to Pakistan.

That night, we slept in the Shabnam Hotel in Khaliq's room. Winter nights were so cold and the hotel lacked sufficient bedding for us all, so it was normal for everyone to sleep with all our clothes worn, even our shoes. When we all put on extra layers of clothing in preparation for sleep, my brother looked puzzled and asked, "Where you are all going at this time of the night?" We all laughed and I said, "We aren't going anywhere. We're just getting ready to sleep." As the Afghan custom of hospitality dictates, s the guys gave the best place to my brother to sleep, the rest of us went to sleep with our shoes and coats on.

Ashraf younger brother

CHAPTER TWENTY TWO

I described my nomadic situation to Farid and asked for his opinion. He said that without money, living in Rawalpindi would be impossible and that there is no work. In Peshawar, there is at least a hope of finding a job. He suggested we go to Peshawar.

Many years ago, someone brought to the attention of my grandmother the plight of two brothers and one sister. These three siblings had neither parents nor relatives to take care of them. My younger uncle took the sister to Kunduz to take care of his children. The brothers, Amin and Naim, stayed with us. Amin later became a worker and always travelled from one province to another. If he was in Kabul, he came to pay his respect to my grandmother and the family.

Naim was always in Kabul and grew up among our families. We were closer to Naim than his brother and sister. Naim suffered from a chronic severe dry cough. He was coughing so much that one would think all the internal organs of his body would be ejected outward along with his teeth. He used to get extremely uncomfortable when he would go through phases of getting red hot, unable to speak, breathing heavily, jumping like a ball from his place, and because of that, we gave him the nickname Naim Cough. The height and face of Naim was similar to the Hindi actor. Randhir Kapoor. The girls and boys of the family called him Rehinder Kapoor.

Naim lived in Peshawar. He was the personal chauffeur of Sayed Isaq Gailani and his wife Fatomah. The Gailanis (Sayed Ahmad Gailani was the founder of the National Islamic front of Afghanistan) lived in the Kababian area of Peshawar. Sayed Ahmad Gailani had a mansion and within two miles was the home of his nephew, Isaq Gailani, and a number of dormitories buildings belonging to Gailani's National Party of Afghanistan. In front of Isaq's house, The National Islamic party had rented or bought several three-room houses. There were several houses. In one of these houses lived the sister of Ahmad Shah (famous singer) and Timor Shah Hassan (famous actor). Her husband may have been a member of the National Party. In another courtyard, in two rooms lived Mujahideen, and in the third room, Naim Cough.

We met with Naim and stayed with him for a while. Naim's behavior

Me my brother Farid and Naim Cough

Randhir Kapoor

was dependent on his mood. Although he was sheltering and feeding us, sometimes he was very hospitable and sometimes extremely ugly and rude. In general, he was socially awkward. One thing that he wished we should do was to pay a devout kind of respect to Isaq Gailani. Whenever Sayed Isaq returned from a trip abroad, he would sit on a grand chair and stretch his hand on the arm. Many people, even elderly folks, used to eagerly wait in long lines to kiss his hand. Naim very much wished that we would kiss the hand of «Agha Sahib», but we saluted him from afar. Naim still had his chronic lengthy coughing attacks. Someone had told him that he should eat or chew the light bulb (in Afghanistan it was called the series light bulb). He had a plastic full of bulbs in his pocket which he'd eat time to time.

While staying with Naim Cough, we were informed that my sister Rohafza was with the Sayed Hakim family. His daughters Zarmina, Amena, and their mother arrived in the city of Karachi and intended to go to Germany from there. Since we had no means of getting there, we just communicated with our sister through letters. I had to prepare food for my big brother. First, I cut the tomatoes and then washed it. One of the mujahedin who also had a cooking job saw me and was laughing at my process of cutting tomatoes. After laughing so hard, he taught me to first wash the tomato before cutting it.

We became friends with Ahmad Shah Hassan and Timor Shah Hassan. These two stayed indoors most of the time, scared to go outside the house because they were threatened and harassed by Islamic parties, especially the Islamic Party of Gulbudin Hekmatyar. Timor Shah stayed in our room almost every day. All day we talked about anything and everything. Timor Shah shared heaps of funny memories from his past. He used to act on stage for several theatres in Kabul. One time, they went to Islamabad with me to escape the claustrophobic atmosphere of Peshawar. We even managed to play music one night, a forbidden act by the Islamic parties of Afghanistan.

Naim, in addition to being Sayed Isaq Gailani's driver, used to say that he works for Sayed Isaq's brother, Seyed Ismail Gailani. He was helping Sayed Ismail sell precious stones and drugs. A smuggler named Aziz used to live in the International Hotel in Peshawar, hence, he was nicknamed Aziz International. Aziz was a close friend of Seyed Ismail. Every day, the devotees of Agha Sahib Sayed Gailani brought goods

Me and Timor Shah Hassan

Me and Ahmad Shah Hassan

Me, Farid, Hayat and Ahmad Shah Hassan

Hayat, me, Ahmad Shah Hassan and Timur Shah Hassan

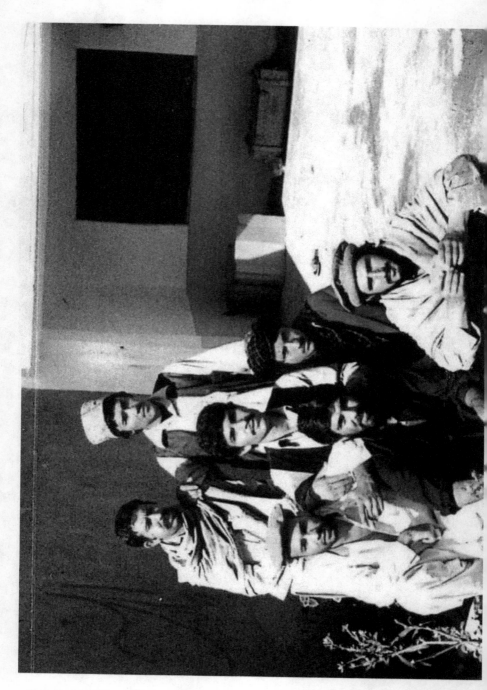

Me with some Mujahideen

from Afghanistan and presented them as offerings to Gailani family. Aziz International's job was to sell these goods to the devotees. He would find buyers and make deals with them. Loading and transporting the goods was in the hands of Naim Cough. Agha Sahib made good money without making his hands dirty or damaging his reputation. There was a wooden ammunition box in Naim's room in which there were layers upon layers of cannabis stored. These sheets of cannabis were as big as the palm of an adult's hand. They were very elegantly and professionally packed with a very thin plastic. This type of cannabis was called raw cannabis and it was not yet ready for sale; it needed to be fortified and 'baked' ready for use. Whenever my Rawalpindi friends, Khaliq, Aziz, and others came to visit me, I would mockingly call to Naim Cough and tell him, "The guys are keen on cannabis, do you have a piece of cannabis for them?" Naim would draw a sheet of cannabis from the drawer and throw it at them. My friends' eyes would pop out as soon as they saw the size of the sheet of cannabis.

After a while, Farid found a job within the Austrian Relief Committee. He worked in the city of Mardan and at the end of the work week, he'd come to Peshawar for the weekends. Sayed Hakim's family was able to go to Germany while my sister remained alone in Karachi. With no other option, she had to join us in Peshawar. Since Naim's room was not an appropriate place for my sister, I decided to take my sister with me to stay with Ashraf's family in Rawalpindi.

CHAPTER TWENTY THREE

My sister and I, together with Ashraf's family, had to rent a bigger house. Farid helped a little with finances, some money came from Germany, and every now and then, we'd borrow from friends. After some time, another brother of Ashraf had joined all of us. Ashraf's family had leftist political views. Each of them believed fervently in their own views and version of communism. Their father believed in old-school communism, his brother was a very sentimental communist, and Ashraf had a romantic view of communism. They had arguments every day. Actually, it's better to say that they had a lot of heated political arguments. Ashraf's mother still not only hid her face from me, but from my sister too, an odd behavior from a mother.

Me and my sister Rohafza

Me Farid and Rohafza

With Gulalai family

Our days and nights were spent in poverty and ill fate. Friends left Pakistan one after another for Western countries. Shabnam Hotel slowly and gradually became vacant from all the departures. Nasir's group, along with Khaliq's roommates, rented an apartment near Islamabad that was on the second floor. Nassir and two other guys were qualified car mechanics. They opened a mechanic garage and body shop to generate their daily income. I occasionally visited them to socialize and catch up on life.

Khaliq's family had sent him the money for getting out of Pakistan. While Khaliq was making enquiries for ways and means to leave Pakistan, one night after returning from the workshop, Khaliq realized that his money has been stolen. Nassir gathered everyone in the room. In a very quiet and friendly manner, he explained this cowardly act to the group and demanded without any violence that if anyone took the money, it should be returned discreetly back to Khaliq without any discussion.

Very soon afterward, the guilty parties were exposed. One of the two had taken the money and went to another city called «Merry» and waited for his accomplice to join him. The accomplice, buckled under the weight of his conscious, admitted to the crime. With his help, the house elders went to Merry City, surprised the escaping thief and took Khaliq's money back from him.

My friend Hayat's father, Janet Khan Gharwal, who was imprisoned in Kabul, was released from prison and came to Pakistan with the rest of his family. They resided in an area of Islamabad called Karachi Company where they rented an immensely large and beautiful house.

My sister always felt extremely uncomfortable with imposing on Ashraf's family. My mother's cousin, Gulalai, along with her two very young sons, one daughter, and husband lived in Islamabad in a second-floor apartment very close to Hayat's family. My sister and I decided to rent one of their rooms. I was very happy because of how close I'd be to Hayat. Gulalai was a good-natured lady, always filled with joy and cheer. Her husband was a very calm man who always appeared sad and melancholy. Their children were fun and mischievous characters. Hayat and I were together almost every day. Since we didn't have enough money to cover our expenses, I started looking for a job. After some time, Hayat and I found

Me and Hayat

Friends departure to USA including Hayat

Abdullah Ali-Ahmadi, Niaz Nazari and Din Mohammad Ali

Niaz, Din Mohammed and uncle Najib

work in a 7-UP factory as the light men. I was working for the first time in my life at the age of seventeen. Our job was to look at empty bottles that passed through a light in front of us. We had to be careful to check and see if the bottle had anything inside and if the bottle was broken. After our checkpoint, the bottle went to the next station to be filled with the 7-UP drink.

The exodus of Afghans to the Western countries had increased. Zabi Gottak, Naim, Zia, Anwar, Sakhi, and many others were all gone. Eventually, Hayat Gharwal and several members of his family also left Pakistan to US. His parents stayed behind, but sadly a couple of weeks after Hayat went to the United States, two motorcyclists assassinated his father when he was going home after morning prayers. There were rumors that the assassins belonged to the Hizb-e-Islami Gulboddin Hekmatyar's. At that time, the Hizb-e-Islami, in addition to having prisons, brutally killed intellectuals and influencers from Kabul and major Afghan cities in public. Janat Khan Gharwal was a supporter of King Zahir Shah. Eventually, Gulalai and his family also immigrated to America. My sister Rohafza and I had no choice but to go back to Peshawar.

CHAPTER TWENTY FOUR

Back in Peshawar in an area called Takaal, we rented one room in a two-room dwelling. The other room was taken by a brother and sister as well. The brother and sister were very good friends. They were always jolly and smiling; they didn't allow the problems of life to bother them a bit. The sister worked with the Austrian Relief Committee. Every month they repeatedly attempted to go to India, but did not succeed. They'd say their goodbyes to us and leave only to return after a few days. Farid would come visit us in Peshawar. One time, he brought a new friend named Najib, who became a good friend to us all. Najib was later renowned as uncle Najib, not because of his age, but because the nickname uncle best matched his look and behavior. Uncle Najib was conscientious, reliable, modest, honest, a person of faith and good friend at any time. Uncle Najib was a very innocent person and always had a warm and kind smile on his face.

Soon after we were in Peshawar, our beloved Mother came to visit

s. We kissed her hands and feet with great gratitude. I was especially pleased to see her after almost two years of feeling her absence every second. When I saw her, I hugged her so tightly and wouldn't let go for a long time, wishing she'd never go away from me again. She was kissing my face and my head, crying, and thanked Allah for seeing me alive. At that moment, I felt the divine presence of the mother and knew that the power of the mother was stronger than the seven heavens. I felt what fortune it truly is to be in the presence of this creature that is undoubtedly a manifestation of absolute creation and love.

"God learned love from mother, otherwise,

the creation was a poem not worthy of singing."

- Kawa Shafaq

Our mother was with us for two weeks. I was calm in her presence in that house, both mentally and physically. I slept peacefully like a baby at night without a fear at her feet. Now, in Kabul, only my father was left with my five-years-old nephew. My mother was worried for them. Eventually, the day came that we had to be separated again from each other. My mother and sister returned to the homeland[13].

With the departure of my mother and sister on one hand and my precarious situation in Pakistan on the other hand, it felt very difficult for me to be calm and even breathe easily. I was beginning to explode with depression and loneliness. Sometimes I went to The Green Hotel to see a few friends that still lived in Peshawar. In one of my visits, a rumor was spreading that a Commander from Kabul by the name Maleem Rahim of Jamaat-e-Islami party is creating a group for Jihad. Since I had enough of Pakistan with its heat and its miseries, I told people at the Green Hotel if such a group is created, then put my name on that list. I knew I could easily pass as a Mujahid because I learnt to perform all the five prayers completely and skillfully. I was no longer afraid of the mosque and Mullahs. Thankfully, my role as a Mujahid never materialized.

In one of my trips to visit my friends in Rawalpindi, I met one of Khaliq's

---

[13] I did not see my mother and sister for about ten years.

Me my mother and Rohafza

good friend, Mama Safi, who had just arrived from Iran. They told me that Mama Safi was arrested in Iran with his friend Niaz for allegedly dealing drugs. The truth is that they smoked hashish and offered it to their friends too. One of their friends had a grudge against the two and reported them. When the guards visited their room, they found hashish. Being a generous and kind person, Niaz wished to save his best friend Mama Safi, therefore he said he was solely responsible for the hashish. Thus, Mama Safi was saved and Niaz was imprisoned for an indefinite period of time.

While visiting Rawalpindi, Mama Safi spoke of a rumor that the Indian government, in partnership with the United Nations, is accepting Afghans as refugees. I went to Nassir's house to find out whether this was true or not. He reassured me that the rumor was true and that his friends Wakil and Assad Jani were involved in helping Afghans obtain fake passports from different countries. Soon, every opportunist was involved in the game and there were all types of people, big and small, experienced and novice smugglers involved in this trade with a very busy market to serve.

I asked for my brother Farid's opinion about going to India. Many years ago, when my uncle was an employee of the Ministry of Commerce in Peshawar, he took my brother Farid with him to Peshawar. Farid had fond memories from Peshawar and its environment was still familiar to him. He told me that he was happy with his work and life in Peshawar. "I'm hoping that the war in Afghanistan will end soon and I can come back home," he said. I was hugely tempted with the opportunity to leave Pakistan for India, so I contacted Khaliq and asked him to find out about the price of going to India. We found out that purchasing a passport, making a photo change and a fake visa was in the range of two-to-three-hundred dollars, although this price was a great deal for anyone and exclusive to us because our connection as friends. As a last hope, I Immediately sent a telegram to my sister in Germany asking for some money. To my great fortune, my sister and her husband sent me about five hundred dollars at their earliest opportunity. I said my farewells and separated from Farid and Uncle Najib. I went to see Khaliq in Rawalpindi with the hope of going to India.

Once there, I found out that Mama Safi's friend Niaz managed to somehow escape from Iranian prison and was in Rawalpindi. Of all my friends, only Khaliq, Aziz and I were determined to go to India. We bought our passports from Wakil and Assad Jani. Our "passports" were the German government's blue travel document, which was issued by the German government for refugees. This was printed in Pakistan and it looked so obvious, even a child could tell that it was a fake document.

Our travel arrangements were finalized. I spent the last night at Nassir's house. After dinner, Abdullah Ali Ahmadi, the uncle of uncle Najib, asked me if I wanted to go out for a walk. "Yes," I replied. I thought it was a good excuse to go out for a cigarette. Among the group, Abdullah was regarded as a wise and kind person that everyone went to for advice. His advice was always sound and objective. That evening, he advised me so eloquently and fraternally that I will never forget it for the rest of my life. Abdullah made it clear to me that vigilance and awakening in such journeys are vital, and that mistimed enjoyments can cause human destruction. He put a lot of these wise words in my mind. He told me to go with somebody to the sea, somebody who wants to be a friend and partner, not just a boat companion. Maybe at the time I was so ignorant and young that I couldn't understand the full value of his advice and thank Abdullah for it then, but today I say thank you dear Abdullah.

CHAPTER TWENTY FIVE

The next day, Khaliq, Aziz and I separated from our friends and arrived in Lahore with sights of reaching India, the land of wonders. The hub of all smuggling activities was a place called the Shabistan Restaurant. The place acted as a de facto Afghan Community Centre where one could get all types of first-hand news and information. The smugglers, including Farid Reesh, Wakil, Haji Shalgham, Aziz International, Haji Sajawol, Qader Diplomat, Aziz e Atash, Hamid Irani, Sayed Amin, and dozens of smaller smugglers came to this restaurant all the time. In the evenings, almost everyone could be found here. Smugglers took turns to send their people to India. At the Shabistan Restaurant, all the smugglers had their turns determined for when they'd send several people to India the next day. We found that going to India was not as simple as we thought. Crossing the Pakistani border to India was open intermittently. Each trafficker would send three to four people a day to India and then they had to wait until their turn would come again.

We spent the second day struggling to find a smuggler to put our name on his list. Hopelessly, we frequented the Shabistan Restaurant and every time we returned; we were met without success. We also found out that well-known and well-connected smugglers were too expensive for us. I kept on hearing that the most famous and "Godfather-type" smuggler was a man called Sayed Amin. He was almost referred to as some kind of celebrity. We never met this elusive guy.

One evening, we were sitting hopelessly in the Shabistan Restaurant hoping that we'd find a trafficker to put us on his list, but to no avail. Suddenly, the restaurant door opened and in walked an obese, white-faced middle-aged man accompanied by six bulky well-built men. At first glance, my heart sunk. I knew him. He was Sayed Ali Agha, a neighbor and friend of my childhood. We have not seen eye-to-eye for many years. He was known by the nickname of "Ali-Agha Donba[14]". He was chubby and his backside was fat. Back home, there was a feud between our two families involving my elder brother Zalmay and his sister. The story is that my brother and his sister were madly in love with each other, but her father didn't approve of it.

Before the war, before the Russians invaded Afghanistan, Kabul was a lively city opening up to international influences and taking giant leaps towards modernization. The city parks were lively places where young men and women met. Parties, picnics, cinema, and music were amongst the most favorite interests and activities of the youth. Kabul and the big cities in Afghanistan were evolving and embracing modernity. The streets were the place of gathering and exploring the chic and modern ways of living. The big cities had a feeling of being on a permanent holiday. Music and fashion were very popular. The youth listened to national artists like Ahmad Zahir, Zahir Howaida, Hangama, as well as famous international singers such as Kishor Kumar, Muhammad Rafi, Elvis Presley, Tom Jones, the Beatles, and other artists.

People's attitude towards love and relationships between young couples were gradually changing and becoming more acceptable to be seen as normal. But as we all know, changing a society cannot happen overnight and It may take several generations to institutionalize this process in the social consciousness, but it had to begin at some point and the process of modernity had begun in Afghanistan before the war. However, it still had

---

[14] Donba means tail.

a long way to go. Every so often modernity would cause rifts and clashes of opinion and values within or between families.

One case in point is my brother and his wife, the sister of Ali Agha. His love story was more heart-rending and difficult than Laila and Majnoon or Romeo and Juliet. This painful love story lasted for nearly ten years with all sorts of grief, distress, and family feuds. One of the prominent reasons was that we were Sunnis and Ali Agha was Shiite. His wife's father was a devout Shiite and had sworn that there'd be no way for these two young lovebirds to be together. After ten years of tears, exchanging letters illustrated with candles, butterflies, bleeding hearts, listening to sad songs, poetry, they decided to take matters in their own hands. They got married privately and without any involvement or the permission of her family. This action poured oil on the fire of Ali Agha's family's anger.

Seeing him, I wanted to hide so he couldn't see me, but it was too late. Our eyes met and he darted towards me, but contrary to my acute fear, he bursted into a smile and hugged me with affection like a long-lost family member. I said, «Oh Ali Agha, how are you? I am so glad to see you!" Immediately, he cut me off and said, «Oh boy! I am not Ali Agha any more. People know me as Sayed Amin and you should also please call me Sayed Amin.» I agreed to call him Sayed Amin. He took me to his table and asked me, "What are you doing here?" I said that we were going to India with my two friends. "Have you arranged your departure with an agent yet?" Sayed Amin referred to smugglers as agents. "Unfortunately, no," I replied. Then, he asked me, "Do you have a passport?" I said, "Yes."

He nodded and said, "Don't worry. I'll manage you and your friends' passage to India." Within an hour, a thin Pakistani man with greasy hair entered the restaurant and came directly to our table. He greeted Sayed Amin obediently and subserviently with two hands and a bowing of his body. Sayed Amin showed him a chair to sit in. The thin greasy-headed man was the Indian-Pakistani border police chief. Sayed Amin asked him which smuggler's turn it would be tomorrow to send people to India. The chief named one of the smugglers. Sayed Amin instructed him with amazing authority and confidence, "Cancel him. These are my family members. Send them tomorrow." The policeman said, "Ok, sir. Ok, sir. Definitely, sir". The policeman told us to see him in the morning before embarking onto the trains. Sayed Amin showed wonderful hospitality and became our angel of salvation.

Later at night, he invited me to accompany him to Hira Mandi. I politely declined, telling him that I once had a bad experience of going to Hira Mandi that was more than enough for my future seven generations. We separated after that; they went to their orgy in Hira Mandi and I went to the hotel to sleep. The next day, all three of us went to the India and Pakistan train station. The border police was already waiting for us. When he saw us, he took our money, valuables, and passports. Then, he took us to the trains. After about 20 minutes, the train entered India, the land of love, art, tolerance, and...the Border Police returned all our money and items to us. He bade us farewell, gave his whole name, and told us, "Whenever you come to Pakistan, remember to visit me." We said our thanks to him, but in my heart, I said to myself that never again would I visit Pakistan and, in the future, if I was on a flight that was meant to fly over Pakistan, I would definitely change that flight.

CHAPTER TWENTY SIX

Khaliq, Aziz and I, with frightened faces in a state of pure excitement, couldn't believe that we were really leaving Pakistan. We were sitting in the train compartment, silently watching other passengers, deep in contemplation. We were unaware of what would become of our future or what would happen next. The train was moving so slowly. I thought it would take hours to get to the beautiful country of India, but quite on the contrary, we arrived at the first railway station approximately 20 minutes later.

Due to our circumstances, and the tremendous amount of fear, from here on to reach the next city of Amritsar, we were to find other safer means of transportation. So, we decided to get off of the train and take a taxi. Coincidently there were two young passengers from Germany who were also planning to go to Amritsar, so we decided to share the ride and the cost between the five of us. Khaliq knew a little English, so he started communicating with them. They asked where we are coming from and where we are going. Khaliq told them, we came here from Germany and now going back. He showed them our German passport that were actually made in Pakistan, one look at our passports, the young gentlemen were so frightened that the first chance they got, they run away. We were so excited none of us could believe that we were truly in Indian soil...that we were FREE!!!

Within those short twenty minutes, Khaliq, Aziz and I entered the world of the culture, etiquette and socializing of traditional India, where every Indian, with respect for one another, puts both hands together and bows and say Namaste and welcomes every one. We entered the Country where the Taj Mahal was located, the symbol of love and affection. It was the romantic gift that was built in 1631 or 1632 by order of King Jahan, the fifth emperor of Gurkhaniand, to love his beloved wife, Mumtaz. It is one of the Seven Wonders of the World.

We entered India, the country where is, unity in diversity, is the word used to describe the culture and customs of India and countries such a those that are extremely rich in terms of culture and customs.

Over the course of three years, I have seen, learned, and got to know the Indians with their unique culture, including languages, religions, dances music, architecture, food, and social customs. Indian customs and traditions are one of the oldest and most unique cultures. I found that in India, there is amazing cultural diversity across the country, with different cultures from the south, north and northeast, and almost every state has its own cultural showcase.

There are few countries in the world that are as diverse and unique as Indian culture. India is a vast country with diverse geographical features diverse climatic conditions, different religions like Hindu, Muslim, Christian Sikh, and a number of unknown religions. It is known as the birthplace of Hinduism and Buddhis, the third and fourth largest religions in the world and all the religions live side by side, in peace and harmony, and so on.

I have witnessed that this country has been an example of tolerance solidarity and non-violence for centuries. The warmth of the relationship and the joy of the festivities set the country apart from other countries in the global fraternity, bringing to life a number of tourists to their vibrant culture that combines religions, festivals, food, art, it attracts crafts dance, music and many other subtle things. Everything, from culture and values to customs, is in the land of special gods, and according to various aspects of rich Indian culture, such as flying kites, cinema, Holi, Diwali, and dozens of other celebrations and festivals.

New Delhi India

Me, New Delhi

Me with friends in New Delhi celebrating the holi

Indians in the day of their liberation, August 15th, fly a ton of colorful kites that make the sky so colorfully beautiful and mesmerizing, so many colorful kites that you could barely see the sky. Holi is a celebration of color, the delicate way of this celebrations, holding different colors, touching the face of a man and woman, indifferent to themselves or strangers, thus congratulating Holi. The Indians are kind and compassionate to the people of love, and they are compassionately kind to themselves and strangers at the Holi celebration of Indian society. In Holi, they make a homemade drink called Bang from milk, almonds, sugar, Cannabis juice, and a few other ingredients that has a completely different effect than any other liquor ever made.

In the same way, I found that there are three empty squares in each area of Delhi. One of these three empty squares is for weddings. The wedding night is set up for guests and the wedding party. (Some nights, we wore stylish and better clothes, so that no one could recognize us, to eat free and tasty food). They celebrate and set up the bars for a week.

The second square is set for the burning of their dead, and the third is for the burning of Satan the Devil (Rawan), who is celebrated by the people of India as the Devil of the Seven head and the enemy of their god, Ram Begwan. The Rawan, which is made of wood and paper, is moved by a Karachi along with countless other crowds. They are playing religious music to the same particular square, and paving the way for connoisseurs of delicious sweets along the way.

Now, we drank freely whenever and wherever we wanted, without any conditions or cultural or religious restrictions. The majority of the Indian community drank libation. In India, three types of liquor were drunk, a completely cheap liquor with interesting colors such as yellow, red, and orange, which the poor used to buy standing in line at Wine Depot. The second was normal liquor, made in India for the middle class, who bought it twice a week from the wine shop, and the third was foreign liquor for the upper middle class, which came in from abroad.

India is said to have 28 states and seven territories, and if I am not mistaken, there is no official language in India, although Hindi is the official language of the state, the Indian constitution officially recognizes 23 official languages. Indians speak a language other than Hindi, and

Bengali, Telugu, Marathi, Tamil and Urdu are some of the other languages spoken in the country. Urdu was a word from Turkish, Persian, Arabic, and local languages in West Asia that eventually became known as Urdu. The language is written in the Persian script and is somewhat similar to the Indian language, which is why Afghans can speak Hindi as soon as possible.

And I found out that family values are highly respected throughout India and are fundamental to everyday life. The structure of the family is patriarchal; a woman must follow her father, husband, and son.

What can I say about the food of the Indian people?

When the Mongol Empire invaded India during the 16th century, it had a significant impact on Indian cuisine. Indian cuisine has been influenced by many other countries. This food culture is due to its high food intake and high consumption of herbs and spices. Cooking also varies from region to region, and wheat, basmati rice, and simple foods with high spice make up India's diet.

Yes, India's foundations of society are extremely religious and cultural. Whether they are poor or rich they follow their culture and religion. For example, the poor individual who even does not have a home and a shelter sleeps on the streets, then from early morning till late at night works hard. Even though he/she does not make enough, but is very content with what he or she makes, divides his earning into three parts one part of which is devoted to food, the other to worship, and the third part spent on movies.

Yes, another passion of Indian culture is cinema. Each state makes a film in their own language, but the strongest is in the Hindi language, then in South Indian and Punjabi. The biggest entertainment in Indian society is film and cinema. Hundreds of thousands of young men and women leave their homes from various cities to flock to Bombay, or Mumbai, for the hope of taking part in this craft to become famous. Unfortunately, no one knows what happens to a lot of them; they could be dead, or just disappeared or become poor.

The top mafias are in the movie business, owning the industry. While the poor Indian who works hard all day, holding the bricks with their teeth, climbing and taking one by one of those bricks to the 10th floor in the

Germany Köln, me my brother Zabi, Satar a friend

Hamburg

Germany Lübeck, with my Classmates

me, Bari and Zabi

Germany 1989

Me with my friend Bashir. Two days after the fall of the Berlin Wall. Russian military uniform

hot burning sun, day and night, or sell water and or give a ride to the fat people from one address to the other in his bicycle and at the end of the day they would pay half of their wages to the film and cinema merchants.

Needless to say, although mafia politics in Indian society have ultimately fostered class influence, people have not changed their way of life in spite of all the problems and difficulties in society, that is, they are extremely attached to their religion and culture. In addition to the thousands of languages, you can also find thousands of religious beliefs, from mice to rats to elephants to monkeys to cows and many more. But the interesting part is that no god of a religion is more beautiful than the god of others, so despite all these gods, they live a peaceful brotherly life.

After three years of living in India I went to Germany. I learned honesty from the German society, learned responsibility, learned respect and family solidarity, learned fashion, learned punctuality, learned respect for women, I learned to respect the hard work of the worker, and learned respect for others.

After sixteen years of living in Germany, I went to California, specifically Los Angeles in California. In America, I felt free and safe. I felt the existence of every race, religion ethnicity living all together in peace, I felt the most beautiful beaches, I felt bright air and bright sunshine, I saw the beauty of sunrise and sunset, I felt sixteen hours of hard work seven days a week, I felt insomnia, I felt the injustice of doctors to treat patients, I felt the smell of barbecue every day, I met sports enthusiasts every day everywhere, I felt the lack of knowledge of society in political, geographical and historical information, I felt the suffering of loan and credit card holders, etc.

In 2004, I returned to my dear homeland after the horrific Taliban were driven away by NATO forces led by the U.S.A. I faced a country that was foreign to me in many aspects. The beautiful city of Kabul clearly reflected the utter ruin and misery caused by the relentless fighting of past years. The people and their interaction and behavior was totally foreign to me. The influence of Iranian, Pakistani and, to some extent, Arabic culture could be strongly felt in many areas, even in government offices and the media. Despite suffering from over-population, filthiness, poverty, corruption, unemployment, ignorance, drug addiction, assault on females, and, especially, air pollution, the residents were so happy that the tyrannical rule of the Democratic People's Party and, after them, their bastard successors the Mujahedeen and Taliban had come to an end,

USA CA me Bari and Zabi

Picnic with the family and friends(Afghanistan 2004)

Afghanistan 2004

me and my wife Roya Afghanistan 2004

me and Bari Afghanistan 2004

and they had instead found freedom, so much so, that the joy was written all over their faces and bodies, and everyone was figuring out how to best use this newfound freedom. The biggest change for the good was the fact that universities and schools had reopened again, which was especially fortunate and necessary for all the little and young girls who had been deprived of the benefit and blessings of education and learning for years.

Since I sincerely loved my home country and its people, I was utterly enjoying every minute of being back there, and, at the same time, was tormented by all the negative changes that had occurred since my departure. While I was there, I came to know and fell in love with a beautiful, kind and educated girl that belonged to an honorable and respectful family. In seemingly no time, we were engaged to marry, and I felt like the luckiest person on earth again.

It was during this time that my older brother, Bari Rahmani, decided to visit and return to Kabul after many years from Germany, and that made our stay there even more enjoyable. My brother, who is also a passionate lover of our beloved Afghanistan and its people, suggested that we see as much as possible of our country while there, and I and my new family liked his suggestion. As such, we had a few very interesting trips across Afghanistan such that an account of each one of them would be too long. However, it must be said with complete conviction that Afghanistan with its ancient history, beautiful, fabulous, and at times even mind-boggling landscapes and scenery, delicious foods and fruits, and utterly interesting people, is one of the most wondrous parts of the world.

However, it was very regrettable that most of these positive and valuable aspects of Afghanistan had almost entirely faded and not even that limited freedom remained unscathed, and, once again and due to the weakness of the current government, various bloodthirsty elements had resumed their criminal activities, especially their shameless assault on the female section of Afghan society.

Today, I am asked after living in Afghanistan for 16 years, 2 years in Pakistan, 3 years in India, 16 years in Germany, and 19 years in the United States which of these Countries I favor the most? And my answer is: The best, and most beautiful period of my life were the three years I spent in India.

AUTHOR'S NOTE

I wrote the truth... the truth that is not a subjective or supernatural phenomenon... the truth that is a clear understanding of theorems and relationships. Ugly and beautiful, they are both parts of the truth. Reality and truth cannot be separated; the two are interconnected.

Through writing these bitter and sweet memories into From Kabul to Peshawar, I never intended to make myself a Ronan and a hero, or to be the only person who has become a warrior. My hope-petals has been plundered by the tragedy of the war merchants. So many died in this business of war. I thought of this fact and wrote this book for two main reasons.

Firstly, because these sad and meaningful stories are part of our life's history, I feel that all survivors must let our future generations know about us and our lives! The trauma is still so painful and heartbreaking, unpausing in all these decades, continuing intensely in our hearts.

Secondly, remembering those days, I wanted to express my heartfelt gratitude to those friends who somehow played a role and helped me survive. Through my emotional memories of the misery, displacement, and helplessness, along with the ups and downs of that era, I got to know the true nature of people. I will never forget the friends who shared their bread with me, who generously gave their time, who whipped my tears of frustration, sadness, and being separated from the warmth of family, comrades, childhood friends, and community. These friends were there for me through the sadness of it all, when I couldn't even breathe. They comforted me and helped me to breathe again. I sought refuge in the warmth of friends' love and compassion because I was a broken bird who fell from the sky. I saw how the roots of those friendships were deeply rooted in the ground, and that in my mind's belief, it looked like the tall pine tree was always green...always spring-like.

I also have bitter experiences of malicious people; whose world is dark and stagnant. This was and is the story of all the united and unjustly oppressed victims of colorful fascism. I truly appreciate your patience in taking the time to read my memoir. With sincere gratitude and in the desire of awakening the nations to achieve social freedom and justice, thank you for your time and patience, dear ones.

Khalil Rahmani

CPSIA information can be obtained
at www.ICGtesting.com
Printed in the USA
FSHW010501270821
84352FS